ARCTURIAN STAR CHRONICLES

VOLUME FOUR

Arcturian
Songs
OF THE
Masters of Light

INTERGALACTIC

SEED MESSAGES FOR

THE PEOPLE OF PLANET EARTH

Arcturian
Songs
OF THE
Masters of Light

A manual to aid in understanding
matters pertaining to
personal and planetary evolution

Patricia L. Pereira

BEYOND
WORDS
Publishing
I   N   C

Beyond Words Publishing, Inc.
20827 N.W. Cornell Road, Suite 500
Hillsboro, Oregon 97124-9808
503-531-8700
1-800-284-9673

Editor: Sue Mann
Proofreader: Marvin Moore
Composition: William H. Brunson Typography Services
Managing editor: Kathy Matthews

Printed in the United States of America
Distributed to the book trade by Publishers Group West

Library of Congress Cataloging-in-Publication Data
Pereira, Patricia L.
    Arcturian songs of the masters of light : Intergalactic seed messages
  for the people of planet earth : a manual to aid in understanding matters
  pertaining to personal and planetary evolution / Patricia L. Pereira.
        p.   cm. − (Arcturian star chronicles ; v. 4)
    Includes bibliographical references.
    ISBN 1-885223-69-2 (pbk.)
    1. Human-alien encounters.   2. Spirit writings.   I. Title.
  II. Series: Pereira, Patricia L.   Arcturian star chronicles ; v. 4.
  BF2050.P458   1999
  133.9′3−dc21   99-27436
                                                                      CIP

The corporate mission of Beyond Words Publishing, Inc.:
  Inspire to Integrity

For my beloved sons, Bruce and Tracy

Table of Contents

PART II

PART III

PART IV

FINALE

Arcturian Star Council, Masters of Light

APPENDIX

Personae

THE ARCTURIANS: Fifth- and sixth-dimensional beings of light from the Arcturian star system (also called the Boötes system). The handle of the Big Dipper points directly to a bright star, which is Arcturus. One of the "jobs" of the Arcturian star council is to serve Spirit on the Supreme Hierarchical Council for Planetary Ascension, System Sol.

CHRIST ESSENCE, OR SANANDA: The Christ, or High Christ Consciousness, the Dove. Higher vibrational identification given to Earth's Avatar Supreme by the Intergalactic Brotherhood of Light. Incarnated upon Earth as Jesus, Krishna, Buddha, Mohammed, and many more. Masculine vibration.

KUAN YIN OF IMMACULATE LIGHT (QUAN YIN): Divine Mother. Bodhisattva of compassion and mercy. Chinese Buddhism.

MALANTOR: A fifth- and sixth-dimensional Arcturian being of light. Malantor means "creator of melodious lyrics."

Arcturian poet and intergalactic counselor-teacher on Earth assignment. Masculine vibration.

PALPAE: Arcturian galactic envoy, Ambassador of Light, Love, and Peace, Intergalactic Brotherhood of Light, to people of planet Earth. Preparer and sealer of documents. Patricia's primary spiritual guardian. Masculine vibration.

SANAT KUMARA: Prince of Light. Brother essence to Sananda. Of the Light Brotherhood Order of Melchizedek, one of the primary orders of the light brotherhoods. Serves on the Universal Oneness Council of Twenty-four Elders.

TASHABA: Catlike entity from the Sirian star system. Serves upon zoo, or ark ship. Gathers DNA and spirit essence of Earth's endangered and extinct species for future retrieval upon Earth or for return to stars of origin. Feminine vibration.

TITANIC BEINGS OF THE GALACTIC CORE: Securers of galactic stargrids to Source Energy. They come from Beyond the Beyond—the regions of Celestial Central—existing both outside and within dimensional octaves, emanating Divine Light of Universal Sun Central.

Patricia's Prayer

May that which comes come swiftly.

May the waiting become that which is sought.

May I, in the glory of supreme moment, know God.

May all beings throughout the many universes do likewise.

May that which I love, all Creation, bound upward into the
holy magnificence of Supreme Grace and Light.

May it be so forever and ever.

    Sat Nam, Sat Nam, Sat Nam.

May the blessings of Divine Inspiration fill me with Light.

May I come to a place of sacred Oneness in every moment
of every day of living.

May Blessed Ones assist me in making it so.

May that which I do serve to inspire others to their greatest good.

In all things may I bring light into the world.

I ask this in the blessings of this moment.

    Sat Nam, Sat Nam, Sat Nam.

In cooperative alignment with all palatial dynamic frequency planets, Arcturian Masters of Light maintain themselves in a state of constant serenity with Absolute Divine Essence. They are dispensers of the ray of Immaculate Splendor (Divine Wisdom).

Arcturians are preparers of intensity actualization (a process of injecting blocks of living Light, precursors to life, into noninhabited planets) for developing planetary systems. They are cosmic teachers of self-discipleship for spiritually evolving species preparing for light-body ascension.

Earth, a Christ Planet

Ultra-refined thirteenth-degree supreme avatars, such as Christ Essence, use time-space vectors for interdimensional travel. Supreme avatars are magnificent beings who are "capable" (an almost non-sense word) of sweeping unevolved systems free of the debris that prevents their inhabitants from imbibing nirvanic elixir as do less restrictive vibrational beings. Collaborating directly with Divine Impulse, these rarefied beings are given, by imperative decree, permission to assist star systems, galaxies, and, in some cases, entire universes in upgrading their base vibrations to rapture-resonance levels.

Christ Essence is Magnitude One high-voltage energy. Christed-level beings synthesize and pull ribbons of prime creative matter from the depths of Universal Central Sun and shape them into stellar and planetary form. In a very real sense, supreme avatars are galactic architects, engineers, and construction superintendents.

It is unfortunate that humans do not feel obligated to assume a spiritual attitude that focuses on cosmic energy as thoroughly as they concentrate on the intricacies of personalities. This bizarre approach is responsible for much convoluted philosophical debate, religious mis-understanding, and a rather odd approach to the practice of Universal Law.

Cosmic Christ Essence is the glue that bonds this solar system in harmonious agreement to the rest of the galaxy. His energy is synonymous with what humans anthropo-morphically interpret as an ultimate being who is respon-sible for and in charge of "running the entire show."

Fifth- and sixth-octave beings, such as the Arcturians, sometimes refer to Christ Essence as Sananda. The name Sananda by no means completely describes His functions or state of being but is an effective means of raising human consciousness from a rather limited viewpoint of the nature of Divine Absolute Perfection. In addition to His galactic chores, Christ Essence routinely assumes human form for the upliftment of individuals and society as a whole and to encourage a higher order of intelligent inter-pretation and application of Universal Law on Earth.

Those of you who have chosen the path of spiritual evolution are issued a divine directive to emulate His Essence in new and innovative ways. You are encouraged, in an attitude of delight, to uplift your traditional ego con-straints and consciously emanate perfected Christ Essence Light. As you assimilate this Light into every fiber of your cellular being, you serve Mother Earth by integrating and anchoring radiant solar-state energy into her magnificent body. As you begin approaching life in a joyful mode, the physical shells that surround your energy bodies will reorient and assume an integrated posture ideal for manifesting what is best termed heaven on Earth. Then, as founding members of the Society of the Eagles of the New Dawn, you will join others in heralding the birthing years of an awakened civilization that, in time, will cover all Earth's citizens with a blanket of peaceful repose.

In appreciation of the depths of your hunger for spiritual knowledge and your desire to be of service in these auspicious times, we are assisting you by providing cosmic information that you are free to approach as being of minor or major significance.

Greetings to the Awakening—I

As clouds of cosmic debris provide elements for birthing stars, you, as an awakening human Soul, resemble a puff of interstellar dust full of potential Love-Light matter. Like a magnet, you are attracting masses of interdimensional energies to Earth. You resemble a miniature sun in that, wherever you go, the light that illuminates your auric body shines like a lantern on the darkened portions of Earth's face.

In this dialogue of celestial greeting, you are invited to accompany your star family in celebrating the joyous occasion of your growing sense of spiritual awareness and activated life purpose. Throughout the moments of your busy days, take time to reflect on your expanding knowledge of the starry clusters and the inhabitants thereof. Integrate this thought form as a mantra: You are One with all things, an important constituent component of Creator's great handiwork. As such, you are the very stuff of which eternity is fashioned—universal energy in manifest reality.

Many scan the skies in hopes of a starship sighting. Essentially, this is a waste of time. You are encouraged to intuitively sense the presence of crystalline interdimensional vessels and their occupants as fields of radiant solar energy. We are much nearer than you dare realize. We are identifiable as airy sensations that make the space around you tingle and bounce with vibrant pleasure. Anticipate and acknowledge enthusiastically that the molecules of invisible gases that brush up and flow away from you are permeated with essences of multidimensional light. Subtly, profoundly, intimately, you are never alone. Therefore, we encourage you to practice thoughtful telepathic communication with us. As you allow your inner ear to integrate the sweet tones of your spirit guides, you will become motivated to accept the inner stirrings of dormant Universal Intelligence. As you do so, you will automatically activate the you-I Oneness principle—the essential element of Primary Soul.

Do not be concerned if you are unable to accurately count the stars in the galaxy, express yourself in complex sentences, or formulate mathematical equations in the manner of superbrained beings. On a cosmic scale, nothing of this sort is important. What is of consequence is your stated willingness to commit to a spiritual life and the intent with which you openly and freely put that into practice. With integrity, adopt the precepts of Universal Law and apply them in every aspect of your life. Attempt to love others without expectation that they will love you in return. Endeavor with the best of your ability to practice compassionate love for all sentient beings. Do these things continuously until you become aware that a *merkaba*, a Star of David–shaped force-field of unconditional Love-Light, surrounds you. Establishing a merkaba vehicle around your body effectively protects you from intrusive activities by lower-astral entities. As you learn to love self and others unconditionally, your capacity to absorb highly charged influxes of galactic-level impulses will accelerate exponentially.

Very little of what is suggested in these inter-dimensionally inspired manuals is new in the annals of spiritual literature. All sacred writings document intergalactic teachings in one way or another. The Golden Rule, which is common to all belief systems, states that humans are instructed to live in harmony and peace with one another. Therefore, with enthusiastic persever-ance, allow qualities of unconditional Love to expand inwardly.

As you walk in your city or town, fill your mind with thoughts of ecstasy and, focusing on your third eye, project a laserlike beam of joy's sparkling, transformative essence around the entire planet. With a challenging demeanor, insist that Universal Peace, Love, and Light be reestab-lished on Earth. You are a citizen of the universe, and these treasured things are rightfully yours, but they were long ago stripped from an inattentive humanity by energy-sucking extraterrestrials, outcasts from the peace-loving, light-emitting greater galactic community. Take command! Break the bonds of illusion the Dark Lords have cast over Earth and all who live there. Challenge your captors authoritatively and with certain knowledge that you, as a principal Earth caretaker—an eagle of the new dawn—are quite capable of doing so.

These things, as so stated, are primary Love-Light teachings.

9

GREETINGS TO THE AWAKENING —I

Greetings to the Awakening—II

In the early years of the new dawn, you, as an awakened, committed starseed, will be nurtured with copious amounts of Love by divine beings. Until the new energies are firmly set, know that an aware, clear mind and heart are your primary allies during the years of confusion. In this context, "confusion" may be defined as a thought-provoking, uncomfortable buffer zone among spontaneously occurring events designed to startle complacent humans into taking deliberate action. Seeing moments of personal confusion as spiritually useful will help you overcome aspects of ego resistance and will accelerate your rate of progress to Self-enlightenment.

Though time grows exceedingly short, it is not too late to turn the tide of destructive circumstances that, to outward appearances, are attempting to engulf Earth and her inhabitants. As has always been and will always be, every known circumstance is situated on a road that leads to heaven, and heaven may be reached on the back of any fine steed. You have the right, indeed the responsibility, to arbitrate your case, and that of all Earth's citizens, before the universal court of Divine Intelligence. To prepare yourself for undertaking this auspicious task, you must alleviate the weights of despair, feelings of separation, and generalized hopelessness that tend to bog you down. Your backup crew and spiritual coaches—refined stellar entities representative of the angelic realms and the various brotherhoods of light—are assisting those who are deliberately attempting to re-create and restructure their and Earth's disharmonic polarities to levels of refined light.

Evolving human minds send waves of gentle thought skyward. Here, on *Marigold—City of Lights*, mother starship to the Intergalactic Brotherhood of Light where Regency star council gatherings pertinent to this solar system are held, we pay constant attention to emission levels arising from humanity's disharmonious neural synapses. When overly intense areas of instability occur, starship personnel send soothing rays of multihued healing lights: liquid ambers, purples, bright pinks, hot oranges, and radiant greens. A virtual sonata of sweetly humming perfumed light spills upon Earth from our crystalline interdimensional vehicles.

Every fifth-octave and higher being, whether its home sun is Arcturus, the Pleiades, Sirius, or otherwise, hums in glorious accord with the majestic resonations of the multiformed councils of the Intergalactic Brotherhood. As such, all representative beings are in harmonious agreement with That Which Is One, better described as Christ

Essence—that which emanates glorious resonations as Celestial Sun.

Arcturians in harmonic alliance with Patricia's thought-emanations are predominantly Palpae and Malantor. They are members of the Arcturian Light Brigade, keepers of a midway stargate, a fiery threshold region in the system of the great Arcturian sun. Beings who fashion light from base matter Everlasting Love are invited to pass through Arcturus's holy portals, a starry gateway that opens into the many mansions of the universe's multidimensional worlds. From far-distant rosy Arcturus, magnificent beings of light urge their celestial chariots over the humming starry grids, joining their comrades from many suns to witness the return of their beloved Dove to Earth.

Arcturian councillors coordinate activities associated with the awakening of the Arcturian contingent of starseeds. As such, it is their privilege to teach and to urge upward that segment of humanity whose Soul appointments are karmically connected to the abrupt demise of the ancient planet Cheuel. Arcturian starseeds are acknowledged members of the holy tribe of elohim (see in Part II, "Elohim and Hybrid Elohim"). From the perspective of the Regency star council, these starseeds emit brilliant rays of light. Although we can offer little tangible evidence in third-dimensional reality to support this audacious statement, it is, nevertheless, so.

Starseed implants from Arcturus, Pleiades, Antares, Sirius, Andromeda, and galaxies and universes quite beyond the range of humanity's observational limitations volunteered for repetitive life cycling and cosmic slumber on Earth. Their long and courageous tour of duty nears completion. Their spirit essences hang like delicate fruit, heavily ripe and ready for plucking. The time to harvest and return them to their home stars draws nigh.

Communicate telepathically with upper-octave beings of light. Even if your conscious mind reveals little evidence that you are participating in a two-way conversation, we suggest you meditate with us daily. Concentrating on the third eye, use common creative-visualization techniques to envision sparkling cities and vibrant starports filled with joyous beings preparing to lift off in multidimensional ships to journey to Earth so they, too, may bear witness to and participate in her glorious transition to light.

With your passionate, dedicated perseverance will your Soul Memories awaken. Watch with astonishment as the "miracle" of past and future time merges into perfect now-ness with present time. Immerse your wondering thoughts in enchanted, long-forgotten mysteries; in images of gigantic pyramids, statues, and ponderous rings of glittering stones; in flame-breathing dragons; in whirling, luminescent extrasolar chariots coursing across night skies. Nothing should cause you fear or trepidation when you time travel in such a manner. You are simply engaged in unlocking a portal of personal Memories—a stargate to your Soul's rich storehouse of forgotten lore.

As you meditate, illuminate Earth's fiber-optic-like grids with vast quantities of high-resolution light. The combined efforts of humans participating in Earth-healing techniques comes closer to effecting a complete restructuring of her damaged grids than most can possibly imagine.

Continuing your meditation, reach into realms far beyond Earth and travel the entire Milky Way galaxy. Endeavor to establish one-to-one contact with your multi-starred galactic family. If you are drawn to this book and others like it, it is an indicator that you have reached a point in personal growth when your belief system is open to alternative realities and unlimited perspectives.

Beloved ones, you who deliberately link your evolving minds to that of Earth Mother are, like her, waiting for an auspicious birthing event.

Unevolved spiritually and fundamentally aggressive, humans have historically been denied admission to the vast playground of the stars for many reasons. They have been kept securely locked within a force-field of third-dimensional spatial time. When embodied, they are granted restricted passage within the fourth-dimensional astral levels, which are accessible to them in dreams, meditation, drug-induced states, and advanced shamanic practice. Astral worlds are where humans dwell between incarnations. We commonly refer to the dynamics of humanity's last twelve thousand years as the Age of Limitation.

While energies associated with the end times and the Age of Turning are set in place, humans must expand their knowledge to include a more profound awareness and acceptance of extrasolar star systems and their inhabitants. The citizens of evolved systems, such as Arcturus, the Pleiades, Sirius, Andromeda, and Antares, are not unlike beings who live on Earth in that they are wondrous to behold. Broaden your perspectives, open your hearts and minds, and commit to establishing unconditional Love on Earth in recognition of the inherent beauty and divine nature of all things. Call for a healing of all karmic situations that, from your point of view, may have caused you or others harm or discomfort. In your deliberations, include your renegade extraterrestrial counterparts, such as the Zeta and lower-dimensional Orions, who have long held humans captive to their nefarious will. Get a sense of the universal prerogative: that every Soul is connected to every other Soul. Embrace an appreciation and love for all beings, no matter their form or degree of Light-Love resolution.

Are you prepared to do these things? It is a critical step in your goal to achieve light-body ascension in this lifetime to view every being, every thing, with overflowing heartfelt Love—even that which is frankly evil. Hold no thing in contempt! All beings, no matter the degree of God-Light radiance they emanate, are diverse aspects of Majestic Creation. The properties of light and dark each being contributes is fundamental to the stabilization of lower-octave harmonics.

Greetings to the Awakening—III

It is an honor and a privilege for us to accompany you who have embarked on a magnificent journey of conscious Soul discovery. Like you, we, too, dream of a better Earth to come. We encourage you to refrain from self-doubt and chastisement when succumbing to small ego gratifications and momentary indulgences, for, as a spiritual warrior of high magnitude, your courage, tenacity, perseverance, and determination to evolve will eventually put to rest aeons of debilitating history that have long held you captive.

If it appears we approach you in a condescending manner, we do not mean to do so. To us, you are an aspect of our own Selves. Unlike you, however, we never experience the pain of separation that prevents you from knowing the truth of your relationship in Oneness with multihued upper-dimensional beings. Your place in time and space with galactic star-masters is securely anchored. It is our intent in this Arcturian-inspired essay to greet you with a message that clearly states our eternal appreciation and Love for you.

Until Earth is firmly fixed in constant Light as indicative of upper-octave preludes, the presence of your spiritual guides and telepathic communication with light-body entities will assume slightly dreamy qualities. Your ethereal guardians are integrating your cellular matrices with the light of cosmic consciousness. Echoes of muted universal sound (Aum vibration) enter your tissues and organs as symbolic messengers that permeate your prelife-encoded DNA with unconditional Love.

It is time for humans to break the shackles of loneliness that have anchored them in mental, emotional, and spiritual frustration and torment in the false belief that they are separate from God. It is also imperative that, as an awakening starseed, you do not set yourself apart as more privileged than those who remain in cosmic slumber. The fact that you have been shown and have accepted a broader vision than most implies that you have received a challenge to assume a greater degree of responsibility. You have received a wake-up call to secure the positive-motivated thought forms of a more spiritually mature society on Earth and to envision her inhabitants living in an Eden-like environment in perpetual peace, harmony, and unconditional Love. You have received an urgent assignment from the Regency star council in the name of the Office of the Christ to shower bright rays of hope upon those who live in chronic despair and to maintain a courageous stance when confronted with the anger and frustration of those who dwell in fear. Your most important task is not only to encourage and manifest your dreams and aspirations into physical reality but, in an enlightened manner, to do the same for those who have not yet achieved your level of cosmic clarity.

If you have studied previous manuals in the Arcturian Star Chronicles series, you have undoubtedly reached the conclusion that page after page is saturated with thinly disguised repetitive writings. Patricia refers to these essays as "cosmic flash cards." We find humans are better able to grasp and internalize new concepts when duplication of effort is applied. That which is oft repeated becomes habit—the practice and reality of everyday life.

A Note on Multidimensional Telepaths

Many become discouraged with apparent discrepancies among channels—more accurately, multidimensional telepaths. At times, channels even contradict themselves. To assist in clarification, we offer the following.

It is becoming common for humans to link up multidimensionally. This is something entirely new in fields of human endeavor. Previously, only a few were multilevel proficient. Since transitional year 1987, many have had spontaneous communication with discarnate spirit beings. Some of these beings are advanced and others are not. Those who have become conscious, aware telepaths since 1987 are true pioneers. There are no schools where they can learn the fine art of telepathic communication. Most are self-taught; only a few have the advantage of an experienced human teacher. Communication with spirit beings is not a society-approved activity; initially, most struggle with a sense of unreality and thoughts of insanity. Because of variations in energy frequencies associated with transdimensional overlay, channels are easily side-tracked from performing the necessary functions of physical life. They must carefully monitor themselves and guard against states of chronic imbalance—what Patricia calls "spacey tinkle."

When reading channeled material or attending channeled functions—including the books of this series—you must consider certain factors. No matter how well-meaning or gifted the channel, some elements conspire to distort the originating message. Unless the channel has attained a state of Buddha-Christ masterhood—very few have—the wavy manner in which energy moves through demarcation zones between dimensions makes pristine, definitive thoughts next to impossible for a human to receive. Other factors influencing a channel include experiential life perceptions; level of telepathic ability; star-system vibrational atunement; practical background, interest, and training in science, technology, history, and metaphysics; conscious present and subconscious past-life conditioning; and, most important, related ego factors.

Information irregularities occur when a human telepath inadvertently taps into variables in an intersecting timeline. Even an untrained individual can activate Soul Memories with a brief whisper of thought, and it is even easier for the gifted to overlap one Earth or stellar time grid onto another. (See in Part IV, "Grid Conduits, Emerging Timelines, and Alternative Futures" for the influence of one's star of origin.)

To ascertain personal truth, approach all channeled information with intelligent, open-minded discrimination.

Because truth is relative to each individual, you must ultimately trust that the channel's connecting link to light-
beings is secure. Keep in mind that channeled material
may only seem to appear inaccurate, for what is true for
you may or may not be true for another.

A Challenge to the Ruling Elite

As night turns to day, that which once was is drawing to a close. These are twilight times between ages. The years that precede and follow the dawning of the millennium will bear witness to a simultaneous ending and beginning of the most significant point in human evolution. Ere long, members of the Intergalactic Brotherhood, Masters of Light, and the angelic realms will come before you as physical presences. Our task is to prepare you for that fine day. As starships land upon Earth, there will be cause for universal celebration, joy, and thanksgiving.

Since the 1950s, we have been challenging Earth's ruling elite and military complexes to put aside their insistence that horrific weapons, military control, and skilled diplomatic maneuvers are the only protective defenses humanity has against the threat of extraterrestrial invasion. Even in the face of mounting evidence to the contrary, most of the prominently positioned find it incomprehensible that a sacred approach to the divine nature and the interconnectedness of all things is essential to cultivating a holy alliance with beings from the majestic light-worlds. Presuming loss of influence and personal power, Earth's leaders for the most part are resistant to spiritual activation, which is imperative for dynamic star-council participation and meaningful interaction within the greater galactic community.

Despite the delaying tactics of world leaders, a new-dawn Earth civilization will soon come into being. Eventually, some of the more hardened members of the human family will harmoniously align with beings of light. Then, they, too, will burst forth in song, and naught but love will infuse the contours of their softening hearts.

In the natural sequence of events, a progressively evolving planet would see its inhabitants spiral smoothly upward on the cosmic harmonic scale. But because of exploitive extraterrestrials and their activities in this galactic quadrant, it has been extremely difficult for humans to break the chains of negativity that hold them captive in the lower-dimensional realms. Long have these manipulative beings practiced their wily craft on Earth. Only the most spiritually advanced humans have understood that freedom lies in the inherent power within all beings to transform nonproductive, muted light to that of omnipresent brilliance. Thus, lower-astral extraterrestrial invaders have held a tight grip on Earth and her inhabitants.

Humans are not innately sinful. More accurately, they are simply uninformed, misguided, and cosmically naive. The conscious knowledge they once held, when the original Edens were in place, was stripped away from them when Atlantis slipped into the sea. Ever since, they have been virtually unaware that their focused thought can manifest as activated form on the third-dimensional level and that there are karmic ramifications of such thought. Contemporary humans living in industrialized societies are so caught up in their busy, noisy activities that they pay little, if any, attention to the strident demands of their inner (Soul) voices. They are virtually unaware that their bellowing, undisciplined thoughts are detrimental not only to their physical health and well-being but to the delicate membranes of their various subtle bodies as well.

Awakening humans, strive with all your might to move beyond the smallness of self that your frightened leaders would impose on you. We tell you: You do have the inherent ability to attain glorious light perfection without permission from any other being. Courageously cast radiant light upon all you meet. In this way you will serve as an example to encourage and empower others. Basic to all spiritual teachings is that light has the ability to mold and contour Love's vibrant energies into exquisite form.

In the beginning years of the dawning century, spiritually prepared humans will receive personal invitations to board the starships. As they enter the rainbow-hued portals, each will be greeted by galactic emissaries of exquisite Love. Until that day, we will place our images into your expanding hearts. They will be transported to your pineal glands so your spiritual eyes can become familiar with light-body beings from the stars.

In unison we stand in awe of your tenacity and perseverance to transform yourselves and your planet to light, in the face of seemingly overwhelming odds. To assist you, we have posted ourselves in the regions of your deepest feelings. We patiently await the marvelous moment you discover us residing there. Surely we do not attempt to hide from you, for we quite frankly anticipate your taking notice of our presence within the silence of your innermost thoughts.

No Earth inhabitant is being overlooked. All are invited to attend the festivities when humanity will be introduced to members of the Regency star council. All have been sent a telepathically engraved invitation to attend. For your maturing comprehension of the nature of our beings, understand that participation is not issued only to a limited number of upper-class, well-placed personages but is fundamental to the entire spectrum of humanity.

To prepare yourself as to date, time, and other particulars for the landing of starships, periodically check in with your higher-Self guardian. The intuitive stirrings of your developing "psychic hotline" will keep you updated. Take steps to consciously align yourself with your personal solar angel, and adjust your life accordingly to assimilate its telepathic suggestions.

We encourage Earth leaders at every level to open their secret files and uncover their hidden agendas so that all are granted easy access to the so-called privileged powers they prefer to keep to themselves. Those who assume it is their right to manipulate information and dispense it when and where they deem advisable are unappreciative of the awesome force of humble individuals who, with a simple outpouring of compassionate Love, thrust megawatt light onto the planet. It behooves those who fancy themselves "in charge" to be practitioners of high ethics and to set moral examples of peace, tolerance, equity, magnanimity, patience, and forgiveness for others to emulate. These simple concepts are basic common denominators of all spiritual traditions.

World leaders, is it not the height of foolishness to pursue a course of downward-spiraling energies that can lead only to entire species destruction? From where and from whom did you receive authority to make such important decisions for all Earth's inhabitants? It would appear you have placed yourselves and everyone else on a suicidal course, for the proliferation of nuclear armaments, unrelenting environmental decay, and social breakdown are quite sufficient to finish the job. Yet you scurry along with abandon while the common folk, who would prefer moonbeams in their pockets, cry out for relief. We are aware much is being asked of you, but you must seriously consider the alternative.

Surely you have noticed that proposals set forth by Earth's Spiritual Hierarchy via the auspices of the Regency star council are not contained in secret documents for naught but heads of state to peruse. Our negotiations are with all the people. Our recommendations are based on Universal Law, precepts basic to glorious omniscient Love and omnipotent, omnipresent Light. We encourage all humans to embed principles of cosmic ethics into the deepest layers of their psyches.

As you reply in the affirmative to galactic contractual negotiations, the stars are yours. Collective as well as individual choices will be honored. Beloved unto us, you are running out of time to respond to the Regency star council's proposals.

Transmitted via the holiest of holies, that which the awakening are learning to appreciate as East-West Oneness or Buddha-Christ Energy, Sananda, this essay is in compliance with galactic contract specifications as proclaimed by Arcturian envoys who bear the privilege of announcing the Regency star council's terms to Earth inhabitants. Galactic participation is not limited to humans, and its broad parameters are being telepathically transmitted to all planetary life forms.

Meditation on Birth, Life, and Death

As you meditate, observe yourself entering a heavily wooded forest. It is richly carpeted with needles and cones that have dropped from hearty evergreens. In the midst of the forest stands a copse of ancient maples. It is early summer. Verdant, softly contoured leaves crowning the old maples dance in shafts of sunlight. Though the trunks and branches are deeply gnarled with the passing of many seasons, the leaves remain fresh and moist, recently birthed in the warming days of spring.

In the quiescent heat of summer's long, lazy days, life is vital yet easily lived. Then, as autumn approaches, an aura of expectancy overcomes the leafy cousins of the pines that monopolize the woods.

Spring's youth realized, summer's passion fulfilled, the leaves succumb to life's sweet impermanence. Enjoying one last ecstatic fling, they dress themselves in brilliant hues. Then, tired and longing to rest, they tug on their stems, straining to release themselves from life's illusive hold. Finally, succumbing to time's beckoning call, they float, serenely silent, to their final resting place upon the forest floor.

One bright, lonely leaf remains tenaciously dangling from a limb. One little soul refuses to detach itself from life's melodious memories. Eventually, however, winter's cold coaxes even the heartiest individualist to follow its companions into the warm world of spirit. Nothing lost, for from the brittle remains of the leafy bodies a fertile compost forms, seedbed material for future forests.

The persistence with which humans hold on to life resembles the determination of the little leaf. Though it is instinctual for third-dimensional beings to hold tightly to the experience of living, ultimately life's tasks are complete. Then, after a time of rest and renewal in the lands of spirit, Soul, all aquiver with excitement and anticipation of life's profound opportunities, readies itself for rebirth. Each incarnation, then, readies Soul for another season in which to grow.

Birth, life, death, and decay, being essential to third-dimensional life, are like leaves, subject to the transitory nature of the seasons. Soul, however, is that ancient core of wisdom whose roots are deeply buried in the rich soil of Eternal Presence. Like a leaf, you may turn with the slightest breeze, for the human is barely conscious of the light-formed spirit body upon which its physical body is formed. Nevertheless, that part of greater Self—Soul—like the trunk of a healthy tree, is strong and firmly planted.

Many humans live apprehensive, stress-filled lives, for they view themselves as impermanent and insubstantial as leaves. They fail to comprehend the reality that they are Souls having transitory human experiences. Attempting to shield themselves from overwhelming anxiety and fear, they cover themselves with coatings of addictive self-indulgences, which, of course, does little to serve their highest good or their Souls' purpose for incarnating into human forms.

A tree is a remarkably effective symbol for visualizing the four seasons of life—childhood, adolescence/young adult, adulthood, and maturity. Even so, it can be argued that the stoutest oak is an impermanent third-dimensional fixture. Soul, being godlike in quality, however, is formed from everlasting substance. Delightful ones, you are Soul! It is only your fleeting lives that resemble leaves: beautiful, fragile, and brief. Struggle to unburden yourselves from the parasitic vines of self-serving habits that keep you bound in their grip. Free yourselves of such unnecessary encumbrances! Freshly liberated, you will mingle as One with light-entities from the stars.

Eventually, all beings emerge as light as butterflies from death's cocoonlike shroud. One of our purposes is to establish meaningful communication with all fledgling starseeds and awaken them by assisting in elevating their energy fields to heightened states of cosmic awareness. To co-create with us in the marvelous task of preparing for light-body ascension, we suggest the following: Study and practice yogic breathing and routine meditation. Open your crown chakras so upper-level light may be thoroughly disseminated throughout your subtle and anatomical bodies.

Listen with the Intuitive Ear

Throw caution to the wind! Focus on maintaining a solid connection with All That Is and radiate Light and Love wherever you go. With the uncensored enthusiasm that is common to children, trust you will manifest everything you need to accomplish your life's highest purpose.

Have courage, my friend. Remember, you are a precious child of the universe. If and when you find yourself bogged down in lower ego's stagnant energies, pull your attention upward and observe yourself from a clearer perspective. Learn to listen to the directions you are constantly receiving through your intuitive ear. Reprogram your ego-brain and train it to run on a perpetual diet of blissful, ecstatic thought. Life is meant to be enjoyed. Approach it with gusto!

Have you made it your primary priority to function from a place of promoting your Soul's purpose for incarnating on Earth? If so, the intricacies of karma and free will, combined with the influx of cosmic energies associated with planetary ascension, make it mandatory that you stay inwardly alert, flexible, and ever watchful of synchronously timed events. We cannot overstate the importance of paying attention to what the inner ear picks up.

It is possible that a shift in Earth's magnetic poles could take place with little or no warning. This possibility, combined with the rapidity with which third-dimensional reality is becoming fourth-dimensional reality, makes it imperative to listen inward carefully. Only your inner, intuitive ear is capable of accessing the moment-to-moment information that higher Self is constantly transmitting.

Earth is poised tensely. Pockets of unresolved negativity are strung about the planet like beads on a necklace—rigid, unyielding thought forms that threaten the vibrancy of the planetary grids. This places stress on the stability of Earth's spatial trajectory and is a major concern for starfleet personnel who monitor volatile planetary-grid connectors.

It goes without saying that Prime Star-Maker is the ultimate authority in these matters. Therefore, during these transformative times, it is your responsibility to establish meaningful communication with It. When directly invoking beings of light (Star-Maker's first-line assistants), clearly state your reasons for doing so. It is your right, indeed your responsibility, to invoke beings of light and request assistance or information whenever you wish, particularly when difficult situations sap your energy.

If you are intent on co-creating with galactic beings in their goal of establishing Earth as a light-emitting planet, you automatically assume the vital task of anchoring light onto the planetary grids. Like a cosmic Johnny Appleseed, you must plant and nurture seeds of encouragement and joy wherever you go. Then, like Johnny Appleseed, move onward without a backward glance, trusting that what you have sown has done so in fertile soil.

We are aware of the courage and dedication to purpose these difficult times require of you. We know of your struggles and how, from a third-dimensional perspective, it appears that very little has been accomplished in restoring environmental vitality and establishing lasting peace on Earth. Yet we encourage you to persevere. Know that every joyous, love-saturated thought you radiate is never wasted but, in fact, is an essential component in energizing planetary and stellar grids with transformative light.

Blessed ones, as you commit to your journey of Self- and cosmic realization, a mutually agreed upon communication access between yourself and refined light-resonation beings is automatically established. Multidimensional angel-like entities are constantly available to you. Contact between humans and their light-realm counterparts is executed by sending thought along the weblike spatial conduits. It is a simple matter to access a spatial "telephone" linkup. All you need do is concentrate on opening your crown chakra to receive and transmit light. Parameters for thought exchange are thus set in motion. Our end of the receiver is kept constantly open. It is up to you, however, to pick up the phone and make the call.

Like unto a Garden

Like unto a garden are the awakening starseeds! They flower and blossom, and violet is the predominant hue they project. Occasionally, even those who are essentially Self-realized succumb to a discordant seed, and their momentary distraction may call forth an invasion of weeds. But such annoyances can always be uprooted. Natural harmony is easily restored to a carefully tended garden.

Cosmically awakening starseeds are endeavoring to conquer the parasitic emotions and addictive tendencies that have long plagued them. They are steadily gaining in spiritual, mental, and emotional health. Although they often feel tried and tested, like hearty flowers they continue their upward growth, unfolding like beautiful blooms for all the world to see. They are learning to establish healthy boundaries that discourage obnoxious intruders. Though intrusions into starseeds' energy fields may take place from time to time, they are not so overwhelming that starseeds' budding vibrancy is permanently squashed. As do all conscientious gardeners, starseeds make sure that the sources of their discontent are identified and quickly weeded out.

Starseeds' every upward thrust not only benefits themselves but also dispenses essential seed nutrients into Earth's hungry body. They are creating an exquisite planetwide tapestry of color. They are carpeting Earth with a vast, vibrant garden that can be observed from far out in space as pulses of brilliant, multicolored lights.

Small groups who periodically meet to dispense seeds of light into the planetary grids are reawakening Earth to her natural state, that of the Garden of Eden. They are planting an ocean of organic material into her light-expanding body. Group activities are certainly encouraged. Those who gather to meditate, heal, share, discuss, and nurture one another and Earth are permeating her nether regions with an expanding ray of brilliant, multihued color. From our position in space, participants sparkle like drops of effervescent light.

Beloved among us, from a fourth-dimensional perspective the garden you are so devotedly planting is already in place. It is only a matter of time before your efforts take root and burst into bloom at your observational level.

Past Lives and Soul Time

There arises in the northern quadrants of the Western Hemisphere energy in human form that once cloaked itself in the guise of Native Americans. Although few retain conscious past-life recall, many individuals now inhabiting Caucasian bodies have had previous lives as Native Americans. With escalating environmental crises, their timeless Memories embedded in genetic codes have begun to surface. They are being drawn deeper and deeper through their subconscious minds into portals of super-conscious clarity, the region where Souls' progression through the stars is recorded.

Those who stir yearn to be done with the prolonged enchantment that has long kept them captive to Earth's gravitational fields. In rediscovering themselves as exquisite Souls, they are learning to focus on a wider interpretation of multidimensional reality. Peering down time's corridors with their expanding perception, they are better able to identify the vast contours of their true Selves. Their logic-loving brains are beginning to realize that the margins of third-dimensional reality are growing increasingly confining. Their hearts yearn to burst through the cloudy veil that lies between them and the upper-level planes. Undaunted by the superficial nature of human society, they intuit that they are permanently linked to Supreme Intelligence through a vast rainbow-hued stellar web. When meditating, their etheric bodies exude increasing amounts of refined Light.

Those who accelerate the rate of their spiritual growth are soon convinced that they are indeed Souls manifesting a succession of temporal bodies. It then becomes easier for them to look on their many Earth lives as a succession of lessons to be learned. They sense that the transient nature of the human condition has no permanent bearing on their Souls' majestic qualities.

Those who willingly step onto the spiritual path and have every intention of traveling its entire course are downloading their Soul files into their brains. Because they are awakening during a period of intense cosmic energy, it is not unusual for them to experience spontaneous past-life flashbacks—not only as humans but also as non-Earth life forms. In meditation they often experience an outpouring of vivid images so real that they eventually conclude they are opening windows into otherworldly realities.

Do not limit your galactic affiliation to one star system. Expand your horizons until you can identify with Arcturus, Antares, the Pleiades, Sirius, Andromeda galaxy, and so forth. Narrow definitions restrict your ability to appreciate self as boundless Soul. Additionally, do not limit yourself by focusing exclusively on only one or two past-life personalities. Determine to re-remember yourself as infinite Soul, as a golden bubble of energetic Light.

There is no need to fret if you find it difficult or even impossible to sense self outside the parameters of your current life. A unique schedule is programmed into the genetic code of every awakening human. Your time will come in due course. It is self-defeating to judge your spiritual progress against others. Ultimately, it is unimportant if you are never able to grasp any other life except the one you are now living. What is critical is that you begin to view self as Soul. As you do, you will begin to shuck off your apprehension of death. When you no longer consider yourself a temporary being, the margins of your life will expand exponentially. For the most part, Souls who choose multiple Earth incarnations have spent time at many galactic ports of call. Indeed, where is it that you think you, as eternal Soul, have been? Indeed, where is it that you think you are going?

Earnestly challenge yourself to prepare to return to the stars. Make ready for an auspicious transition to Light. As a committed spiritual being, you have been provided with a team of interdimensional guidance counselors of unlimited resources. It is their task to help you unlock your cellular Soul Memories so you may reestablish yourself as a spark of immeasurable potential—an awakened citizen of a vast universal complex.

Query your inner teacher(s) with proper questions such as "Who is it I truly am?" and "What is it I have come to do?" Every day, for the rest of your life, commit to expanding your awareness of self as a multidimensional being. Imagine yourself as a beam of light sailing freely from star to star and from galaxy to galaxy. Meditate on the many times you have gamboled freely upon the space grids, and get a sense of their flowery aromas.

Many of you consider yourselves emotionally bruised or physically ugly. Yet in our eyes you are incredibly beautiful. Your stirring spiritual minds send strobes of effervescent light through Earth's darkest regions. Your daily prayers and seemingly small personal conquests guarantee not only your return to a greater reality but also Earth's ascension into the regions of Light. If all the wonders of your Soul's long journey became instantaneously known, however, you would be overwhelmed. You would plunge into a state of spontaneous insanity. Your human neural system is simply not capable of downloading that much information at one time. Thus, you are protected from cosmic overload with pieces of somewhat elusive visionary and telepathic data. In spite of these precautions, much of your grief and suffering is due to your intuitive realization that the majority of your Soul Memories are simply not available to you.

Now that you are being progressively reactivated to take your place as an aware galactic citizen, centuries of karma are releasing their gravitational hold on you. In concert with the tremendous influx of waves of cosmic energy that are pouring over Earth, your genetic matrix is being reprogrammed to integrate additional strands of etheric DNA, thus making it possible for you to access higher-level information.

As an emerging eagle of the new dawn, you are encouraged to think of yourself as a member in good standing of beloved Archangel Michael's troops—as a warrior for the Light. As such, it is critical that you consistently examine the thoughts behind your actions. Ask yourself what your real motives are and to what level of spiritual maturation you are prepared to dedicate your efforts. As you attain a greater understanding of your purpose for incarnating on Earth in these dramatic times, query your inner guides as to what your eternal Soul has borne witness to.

Although you have incarnated into the lower physical dimensions, your Soul resides in universal regions of Infinite Light. Soul is Creation's base fuel—radiant Essence Energy. Soul is unencumbered by spatial-time restrictions. Soul is the golden elixir from which stars are created. Soul is the cosmic "axle grease" upon which galaxies turn.

Self-limited? Ha! Feed your life with Soul-energy gusto. Fuel your tanks with unlimited quantities of Love and Light. Illuminate every inch of ego darkness that lower self is prone to inhabit and habitually protect. Let go and hold tight to your seat! You are being primed to take off on a magnificent adventure. You may think spiritual life is a difficult, exciting, and even dangerous sport. Nevertheless, we tell you, it will be well worth the effort to prepare for the wondrous trip that lies ahead.

Strive to neutralize all unresolved issues. To the Soul the things that plague you are no bigger than fleas on the back of a dog. Lower self feeds and thrives on negative energy. Unless you are extremely advanced spiritually, your personal demons will remain active as long as you hold on to them and associate with tormented, angry individuals. If you give such beings the opportunity, they will surely suck you dry. Feed a healthy dose of compassion and loving light to discomforting emotions and situations. Do not ignore individuals who attempt to manipulate and control you, but as primary companions seek out others who, like yourself, are progressing spiritually. Do everything you can to serve the evolutionary times from a place of awareness. Ready yourself, when the time is right, to catch a multidimensional shuttlecraft Home.

Eden Returned: Plant and Animal Ascension to Light

This transmission is directed from Tashaba of Sirius with Palpae of Arcturus.

Beauteous morning scatters night's shadows as the woods awaken to first light. Emerging from the trees into a broad sunlit meadow, the forest creatures gather as if responding to an urgent summons. As the peaceful assembly grows, it is enhanced by the vibrant essence of each new arrival. Beings of light observe this ritual from their starship almost overcome with an outpouring of love, for they know that the congregation of animals is a symbolic precursor to the moment their beloved human family transmutes the Earth plane.

A song riding upon the wind that only they can hear has called the animals from the safety of the forest. One by one they enter the meadow and take their places in silent expectation. Multidimensionally aware, wild creatures have long anticipated the return of the One who anointed their Souls and birthed them into life.

Hyper-alert, the forest creatures scan the sky for signs of their Dove-like leader. Sooner than they had dared hope, it becomes evident that the sky-blue curtain is parting before a window in space. Joyously, they watch as a fleet of golden orbs emerges through the window, circles, descends, and quietly lands within their midst.

With little fanfare, animals who know their time has come board the ships. Silently, the orbs ascend through the morning mist and rapidly disappear behind the closing portal. The remaining animals, those whose time on Earth has not ended, rapidly disperse into the trees. The forest breathes a sigh as the animals renew their patient vigil. In the meadow, the only evidence of a recent starship visit is a hint of dappled sunlight upon the rocks.

Blessed fortune favors animals and plants whose spiritual natures are unrecognized by technology-loving humans. Earth creatures being squeezed into increasingly narrow living spaces are attempting to telepathically communicate to their two-legged counterparts that the return of the Dove is imminent and that a holy event of this magnitude should be approached with an attitude of innocence and humility. Animals departing Earth on starships reflect beams of unconditional Love upon all sentient beings. They would have you understand that many endangered and extinct Earth species are being reformatted to light-body status. Most wild animals and plants are more cosmically aware than their domineering human cousins, who consider themselves species superior. Thus, it behooves you to consider what truly constitutes a valid indicator of advanced intelligence.

Like humans, animals—and to some extent plants—must assert territorial boundaries, regularly kill other sentient beings to obtain sustenance, and periodically defend life and limb, for these things are necessary to sustain third-dimensional life. Food, water, and spatial requirements are common denominators for all Earth species. Participation in and attitude toward acquiring these life-sustaining commodities is not necessarily a valid marker of unpreparedness for light-body ascension—except in humans.

In every respect, highly placed dimensional beings abide by Universal Law. Universal Law is ab-Soul-lutely just. "Thou shall not kill" is not only spiritual law on Earth but is recognized by all galactically aware citizens as the Divine Will of Omnipresent Being. As previously indicated, the ability to completely fulfill every aspect of Universal Law was taken from you by the Dark Lords, who convinced Earth's inhabitants that the Word as spoken in holy scripture contains clauses that need not be obeyed. Unfortunately, because of the Dark Lords' manipulation, repetitive life-death sequences became Earth's dominant feature, and the limitation of third-dimensional spatial time became necessary for you who are tormented by suffering, hunger, disease, pain, and death.

We suggest you forgive the unfortunate Dark Lords who have long held you captive to Earth's gravitational pull. Truly, they do not know what they do. Choosing to ignore Prime Law, they have successfully manipulated themselves into a cold, heartless situation that knows nothing of unconditional Light-Love. As you let go of the influence of their unsavory doings, you will participate in transmuting individual and collective unresolved karma and assist Earth in raising her energies one more rung on the galactic vibrational ladder.

An interlude with Divine Mother, Kuan Yin of Immaculate Light: Perchance, dear one, you came upon a deer in a forest. The deer, elegant in countenance, bends toward you and allows you to brush its forehead with the palm of your hand. Misty-eyed, you look deep into the liquid depths of the deer's eyes, and you are privileged to see its Soul. Recognizing the level of contact, the startled deer turns and runs into the safety of the forest.

Now you stand quite still so that you can completely appreciate every nuance of the serene environment you find yourself in. It is morning. The forest is alive with activity as many creatures move about in the cool early hours of the day. Above, the leafy canopy reaches for the sun. Below, tucked into the crevices of the forest floor, delicate flowers poke their beautiful faces toward filigrees of light filtering through dancing leaves.

The sublime loveliness of the scene stirs at the lush places in your heart and sends sparks of Soul Memory racing through your mind. Your imagination stretches to accommodate every detail of the mysterious vision.

If your observation is acute, you will catch a brief glimpse of your own Soul, just as you did when you looked deep into the eyes of the deer. If such an encounter with your own true Self leaves you feeling a bit overwhelmed, your mind may quickly withdraw to focus on more mundane matters. For you to see your Soul nakedly is a profound experience with enormous implications. Can you, having glimpsed Self as Divine Energy, intelligently deny Soul's eternal nature?

Concentrating on your inward vision, turn your attention to the animals and plants that make up the dynamics of a simple forest scene. Allow yourself to be transported to an awareness of the interconnectedness of all things—of universal Oneness. In a place of peaceful meditative harmony, know that even the most minute things are sacred and conscious vital beings.

As you complete your inward journey through the rising mists, may the Dove of Peace settle beside you. May the sensation of being one step closer to Home comfort you.

The Arcturian-Atlantis Connection

Long ago, in cooperation with a diverse gathering of multisystem inhabitants, beings from the star system Arcturus participated in establishing an Atlantean imprint on Earth. Atlantis was an experiment that went awry after several generations of success. The greater Arcturian experiment, however, was focused on spiritually aware Lemuria. Our influence was felt even farther back in time prior to the separation of the Continent of One. Be pleasant with this news and try to understand its impact on human history.

In early starseeding times, to avoid thousands of years of darkness in this galactic sector, the combined star councils established a powerful Atlantean Earth grid connecting the regional stargrids, which they thought would override an increasingly negative influx of lower-state beings. Unfortunately, the free will of the predominant Atlantean intelligentsia conspired with those off-world beings to upset star-council efforts to establish a spiritually advanced multidimensional compound on the Atlantean continent. Atlantean elders, who closely resemble current society's cosmically immature leaders, became increasingly distracted by glittering enticements of what is best described as an extraterrestrial "play toy" phenomenon. Enthralled with advanced alien technology, they became overdependent on available planetary resources and began to ignore the stabilizing effect of the Lemurian elders' spiritual teachings. The latter's predictions of impending doom were cast aside as unsubstantiated nonsense.

It should not be assumed from these brief remarks that the Atlanteans were an overtly negative people or that they were uncommonly seduced by their extraterrestrial visitors. In most respects, they were a complex, fiercely independent people.

In the early years of the galactic experiment, Atlantean leaders, establishing free-will prerogative, attempted to resolve their escalating social and environmental problems by blending light images with shadow images. Wisely recognizing the need to equalize polarity imbalances within the physical, auric, emotional, and mental bodies of Atlantean citizens, they erected a countrywide system of healing temples. Many modern-day healers are psychically awakening to Memory images of their lives in Atlantis as temple healers, priests, and priestesses.

The popular notion that a united world-governing body poses a tyrannical threat is a residual Memory of the star council's plans for Atlanteans to attain higher-state at-One-ment that went awry. Strange as it may seem in the face of perceived evidence to the contrary, humanity at the turn of the twenty-first century has come closer to achieving cooperative world harmony than any civilization since the Continent of One was rendered into many.

To expand on your growing knowledge of historical and spiritual matters, you must know that multisystem star councils originally permitted fear-based entities to establish residency on Earth for the purpose of balancing universal harmonics in third-dimensional spatial time. The Lemurian elders who were actively affiliated with the multisystem star councils anticipated that these negative beings would have less deleterious impact. In retrospect, we can see this was not the case. After the fall of Lemuria and Atlantis, the Dark Lords moved en masse to captivate newly emerging human civilizations. Because of their general success in so doing, the exacting nature of spiritual teachings and disciplines has been less understood or practiced by the human collective.

Eventually, ruling Atlanteans plunged their country into chaos. Light-emitting individuals immigrated to other lands before they were ejected or destroyed. A weight of unyielding darkness plunged Atlantis deep into the ocean. The experiment of promoting advanced galactic knowledge in an immature species was over.

Today's human is witness to the use of matter-based technology and the overmanipulation of natural resources unequaled since Atlantis. The star councils have also gained much insight over time. Their primary reason for contacting humans who struggle to awaken from the coma of third-dimensional sleep is to encourage them to anchor light into Earth grids to help offset escalating spiritual inertia.

We anticipate you have received some clarification as to the Arcturian-Atlantean connection. We suggest discernment in your studies of all off-world civilizations. Learn to ascertain the primary nature of all extraterrestrial visitors. An indicator of less-evolved systems is that they tend to be goal-oriented technologically. Off-world beings coming from material-based systems are among those who have historically caused you so much trouble.

Beings of light hailing from the fifth and sixth harmonics or higher abide by principles of manifestation via the use of unconditional Love in concert with the will of Divine Creator. Technology parameters common to light-world beings are established on applied Love-Light essence. Energy in this form is noncorrosive, nonviolent, and nonthreatening to all life.

Possibilities Inherent in the Dream State

Dreams are packets of multidimensional knowledge that symbolically assure dreamers they are connected to and are receiving higher-dimensional information. Dreams promote self-knowledge, a precursor to Self-realization that is a necessary step for those wishing to attain the God-realized state. Self-knowledge is so powerful a force that it has the capacity to change every aspect of one's thinking and to transmute any deleterious karma.

Those who ignore the exquisite, detailed tapestries of their dreams severely limit their capacity to offset multi-level influxes that threaten the stability of their corporeal lives. If they are crippled by attitudes of limitation, they lose the capacity to consciously readjust energies affecting their physical, astral, emotional, mental, and causal bodies and set themselves up for excessive energy drains on their life force.

The integrity of dream-world consciousness propels one effortlessly through gravity-free astral worlds, providing momentary relief to the human mind, which tends to focus on weighty issues promoting stress and anxiety. Those who deny their multilevel aspects often feel incomplete and privately view themselves as insubstantial beings. Sadly ignoring their most effective means for accelerated self-knowledge and activated problem solving, they wallow in emotional chaos and unresolved personal issues.

Those who acknowledge that worldly sensations and life happenstances constitute only one universal level know that dream planets are very real places. Profound as they may appear, astral worlds are only another step into even higher octaves of the universal domain. Humans who realize they are not completely bound to linear-time restrictions are learning to consciously open doorways into alternative realities through dreams and meditative visualization.

Many cannot imagine that there are such things as planets and beings who exist beyond humanity's perceptual range. Those who are cosmically unaware generally believe that Earth is the only place life exists. They simply do not comprehend that the universe is made up of both complex and subtle levels of vibrational energy. If told that dream worlds are in fact real, they would most likely respond, "No, that is quite impossible. After all, you see, dreams are nothing more than fluffy imaginative fantasies, nonexistent playgrounds where rules are off and one can do as one pleases."

Like it or not, dream worlds are genuine, and quite assuredly all rules are on! To overturn any aspect of Universal Law while dreaming has cause-and-effect ramifications as surely as irresponsible behavior does when awake. Divine Law is based on principles of unconditional Love and is vitally active at every galactic level.

Another important aspect of the dream state is that spiritual guides are able to contact and teach their assigned charges. To ignore the advice of and opportunity to interact with advanced beings when sleeping is certainly foolish and self-defeating.

It is said, "It is not wise to fool Mother Nature." Although physically bound to Earth by laws applicable to third-dimensional reality, when traveling in the dream body, humans are subject to laws governing fourth-dimensional astral worlds. Complicating their situation even further, many who live in spirit-stifled societies are so engrossed in the distractions of electronic and mechanical technology that they generally fail to apply standards of Universal Law in both their awake and dream relationships.

It is imperative that you consider the symbolic images of at least your more prominent dreams. Even a rudimentary grasp of their meaning will help you align your energy-body vibrations with Earth's as she moves into her final transformative phases. To accomplish this feat effectively, contemplate all symbolic dream images intuitively. Keep a dream journal to recall your dreams and your spirit-guide dialogues.

To assist your progress in the worthy endeavor of incorporating dream reality into everyday life, many excellent references and teachers are available. Be guided by that which most favorably draws your attention.

Mountain Tragedies

This transmission is directed from Malantor of Arcturus and Tashaba of Sirius.

Mountainous regions, particularly in the last wild stands of Canada and the "United" States, are experiencing escalating loss of human life, primarily sports-related. It will soon be difficult to ignore what is already obvious. Sequential death-following-death occurrences are meant to forcefully drive home the point that humans have allocated too much land surface for living quarters (fourth of a whole) and for recreational activities. This causes much distress to wild plants and animals whose natural habitat is limited to mountainous regions. Spatial allowance for wild creatures has been reduced to a bare minimum. This is particularly deleterious in winter when those creatures' survival stresses are at peak level.

The forests and meadows of mountain expanses increasingly echo from the swoosh of skis and snowboards. Even more bothersome is the clamor of motorized vehicles. In regions set aside by law to protect the last remaining stands of old-growth trees and indigenous wildlife, demands are made for more resource and recreational access.

The lack of leaders to define and uphold the needs of wild animals and plants is reducing available habitat even further. Land allocated exclusively to other species—many of whom hover on the brink of extinction—is marginal. This does not bode well for sustaining the integrity of Earth's natural places.

Patricia, you have asked us to explain the symbolism behind the growing number of recreational deaths and the number of avalanches that seem to be taking so many lives. We say that these events are not symbolic. The mountains and the animals and plants who live there have decided to take action to protect themselves. They wish humans to know that further excessive intrusion into these already overused regions is intolerable to their well-being. They are prepared to "conk a few beans," as it were, in hopes of getting your attention. They believe that intruders will eventually get the message; in light of human propensity to disqualify the needs of other species, however, it does not seem likely in the immediate future, although environmentally conscious people are on the right track.

This is yet another wake-up call summoning humans to make Universal Law their priority, as do all cosmically alert galactic citizens. There is growing concern, not only among humans but in the plant and animal communities as well, that pristine forest and mountain habitats will soon be a thing of the past.

Do you think humanity will get their rather forceful hint in time, Manitu? Sadly, as things stand at this writing, neither do we.

PART II: SANAT KUMARA, MASTER OF LIGHT

―――――――――――――

Speaking on behalf of the Christ Essence, He Who Bathes the World in Light, by the direction of the Holiest of Holies, I Am Sanat Kumara, of the Universal Oneness Council of Twenty-four Elders, dispenser of Divine Light throughout this galactic region. I Am that which is referred to as the Prince of Light, brother to Christ Essence, Sananda. My I Am presence is recognizable as the God-Self that dwells within you.

The Council of Twenty-four Elders comprises beings who blend and bond Divine Light-Sound into birthing and advancing galaxies.

57

ARCTURIAN STAR CHRONICLES

A Challenge to Mature

Intellectuals disregard the possibility of stellar communication via telepathy. Arrogantly, they insist that humans are Creation's highest form, that they are the only intelligent beings in the universe other than perhaps luminescent angels—whose existence is, of course, extremely suspect. Certainly, they say, the chances of life existing outside Earth are minute. Recently, however, with the proliferation of paranormal events and sightings of probable extraterrestrial vehicles, their hard-line stance is beginning to crumble. It would be extremely foolhardy to remain resistant in view of growing evidence to the contrary.

Long a favored discussion inside Earth's behind-the-scenes governments, classified investigative materials remain hot topics. This is also true at most levels of society. Although at least somewhat aware of the facts concerning extraterrestrial interest in Earth, world leaders continue to insist that they do not owe the population a lucid explanation. They privately acknowledge a greater truth: They plan to play out their game of evasion as long as they think they can get away with it. However, their years of attempting to draw a curtain of illusion over the people are drawing to a close. It will not be long before Earth is visited by a formal delegation of light-masters; your leaders' tone on that auspicious day will most assuredly change.

Sacred tablets relating true history are stored in the etheric city of Shambhala in the Gobi Desert. Mirroring Earth's holy celestial documents, all sacred human texts contain references to the star-masters—godlike beings associated with early human mythology. The Bible, the Talmud, the Koran, the Vedas, the collected teachings of the Buddha, the secret doctrines of the Masons—indeed, all sacred literature attests to the reality of the great brother-hoods of light and their historical interactions with humans. Nevertheless, even the most intelligent and well-educated humans fuss over the possibility of a universal cosmology designed around a centralized celestial govern-ing body. What is (subconsciously) frustrating to elite lead-ers and thinkers is that until they learn to conduct their affairs with high degrees of wit and sanity they will fail to hold positions of authority within the celestial body of One. Until then, they will be subject to constant urging "from above."

At its current level of spiritual maturity, humanity displays little intention of wisely and effectively Self-actualizing. The number of so-called in-charge individ-uals who are willing to divest themselves of self-destructive, retaliatory methods of governing can be counted on one hand. Most world leaders, as well as their subjects, demand that an eye be taken for an eye and that they have the "right" to bear arms against other members of the human family.

In this spatial time, the universe in which Earth's sun is located is undergoing a ninth-degree angular spin—an upgrading of its vibrational alignment. A ninth-degree universe is one that has completed a certain level of Divine intention and has heeded a summons from Omnipotent Omnipresence to modify its complex structures to a more resonant hum.

For the most part, humans are not consciously aware of the accelerating surges of cosmic energy that have been sent Earthward since the transformative year 1987. Few know how to use the beneficial aspects of these waves to prepare for light-body ascension. These powerful energies originate in Beyond the Beyond regions at the outer edge of the universe. Locally, a magnetic grid has been developed to draw the Milky Way galaxy a step closer to Celestial Home, the residing place of magnificent titanic beings.

This is all too much to comprehend, we dare say. However predisposed you are to accept or reject these data, it is imperative that you make an effort to overcome resistance to multidimensional references. It is also important to do more than simply gather information for the purpose of satisfying curiosity. To gain an upliftment in vibrational attunement you must incorporate Light-Love into all resistant emotional, mental, cellular, and DNA areas. Modify yourself so positive thoughts and actions based on Universal Love are expressed outwardly; ingest high truth as you would nutrient-high foods. Using creative visualization as a method to spiritually activate and wisdom and discernment as key motivators, manifest a complete reengineering of your cellular structures and your genetic code. This is imperative if you wish to transform your physical body to perpetual radiance. Make an effort to transmute at every level.

The tantalizing aspects of paranormal events are distracting to true seekers of spiritual wisdom and easily dissuade them from concentrating on what is really important. Humans are running out of time to perform basic tasks related to Soul-essential personal and planetary housekeeping. Opportunities to obtain additional incarnations on third-dimensional Earth are almost nil; her time for residing in lower-spectrum light is quickly coming to an end. To live on fourth-dimensional Earth, by necessity you will need to maintain your vibrations in harmony with hers. All who do not will be provided with alternative planetary living space. Personal issues must be attended to, processed, and vibrationally readjusted with diligence and tonal precision. You must pay attention to subtle, insistent, lingering, negative-based thoughts and transform them to higher-resonant levels.

Your principal spiritual guide and attendant subguides are always available to you, but the bulk of effort must be yours. These entities are guides and teachers; it is not their task to do for you what is yours to accomplish.

We are aware that telepathically transmitted extra-stellar materials are confusing at best and that most of you are more or less intimidated by them. Oddly phrased writings that speak of multiple universes, light-body beings, and the ascension process are inexplicable to most humans. Although Earth-based interdimensional telepaths have the ability to organize pieces of cosmic data into loosely accepted rational terms, cultural and historical indoctrination confines the majority to a limited reality somewhat loosely based on observing the results of experiential stimuli. The idea of one-on-one telepathic contact with invisible beings, let alone their insistence that humans are being asked to lay the foundations for an entirely new society using techniques of co-creative visualization and cooperative endeavor, completely eludes them. At their current state of cosmic sophistication, it would never occur to them to align their thoughts with those of beings who live on subtle-plane worlds. To protect their sanity and emotional and intellectual well-being, the masses have entrenched themselves behind an illusionary facade that is quite common to slumbering entities who inhabit third-dimensional planets. Sadly, the conceptionally rigid will not easily accept the notion that they have the ability to consciously ascend into an energetic light-body.

It is not our purpose to tease and torment. However, we do mean to urge you to strive for a greater degree of alertness. We suggest that you consciously evaluate all ego limitations for the purpose of integrating light into your physical body and that you limit your self-indulgences to things which give you a heightened sense of integrity and self-worth—that which is exceedingly good. Constantly be on the lookout for self-defeating thoughts. Consciously emanate laserlike light from the depths of your heart and crown chakras—an activity that is both personally and collectively transformative.

Throughout these essays, and the ones preceding them, we have explained our mission as best we might through Patricia's particular brand of telepathy, use of language, and predisposition to metaphysical philosophy. We have attempted to explain as simply as possible the position of higher-world beings and the manner in which we express Universal Law as applied unconditional Love. We have indicated that on a cosmic scale unconditional Love is force-field-quality Prime Energy equal to and surpassing a star's ability to radiate light.

We trust we have not entangled you in a scattered web of obscurities. A heart-mind reading of these data is the best approach. Your brain's complex electromagnetic neural structures can absorb at a solar transmission frequency for only brief periods without becoming fatigued. This is not meant as a criticism of human intelligence. It should be understood that the physiology of a third-dimensional brain is in danger of neural burnout until such time as the host body has readjusted its cellular and DNA components to a more refined vibration. Individuals exposed without proper preparation to solar intelligences often experience mental breakdown, even insanity. A physical body must be primed before it can sustain lightning-like telepathic pulses. Though your brains are quite capable of communicating with higher-world beings, unduly long sessions can be detrimental to psychological and physical health. We advise fifteen to twenty minutes at onset, with half- to whole-day rest periods. As your vibrations are raised through meditation, diet, and other spiritual endeavors, you can remain in contact with inter-dimensional beings for longer periods of time.

The human heart-mind, being the point of Soul connection, is capable of focusing rays of passionate intention into the ethers. Thus, the pathway to the stars is through your heart. Brain-mind is meant to enhance visual and language components of telepathic relays. The initial pulse, however, is received via Soul connector, where the energy is transferred to the pituitary and pineal glands. This is why your heart and not your intellect is where you truly comprehend your connection to alternative worlds.

Souls residing on Earth, before they are capable of conscious telepathic contact, are provided with multi-dimensional access through visions, dreams, and out-of-body experiences. Because Earth is a primary teaching planet for evolving Souls, humans exposed to cosmic-level information are somewhat like kindergartners finding themselves thrust into schools of advanced learning. Therefore, it is not uncommon for them to become agitated, befuddled, and bewildered by the paranormal or by the approach of light-entities. Issues of free will aside, higher-world beings do not habitually present themselves as physical sightings until their human contacts are prepared for same. Our favored approach is to mirror our light-bodies to you in visions and dreams so that you become familiar with the presentation of ascended beings in preparation for your own graduation to light-status.

Coping during the Transition

Humans, you live in the most fortuitous time in history—the culmination of millions of years of a progressive, Divinely inspired plan for the maturation of the entire universe. It is your generation's task to call the star people to Earth.

All things being One, high energies required for the transformation of Earth are in alignment with the upward swing of the universal chorus. These energies are reflected in the extreme states of anxiety in which humanity carries itself as it moves through a spatial time gate that will eventually open into the domains of an enchanted, golden land.

Your fluctuating levels of stress are directly related to the ease with which you are able to absorb and respond to the incoming solar and cosmic energies necessary for integrating one spatial dimension into another. Currently, Earth is at an overlapping stage between the third and fourth dimensions.

It is our desire to provide you with some practical ways for living in this extraordinary time. Probably most important is to provide yourself with a physical sanctuary for daily retreats into peaceful, restful repose. This can be as simple as a comfortable chair in your home, a special room, a walkway by a river, a tree in a park. It is particularly important to maintain fluidity of being and balance when intense levels of physical, emotional, and mental activity begin to take their toll in the form of frustration and fatigue. Become so finely tuned that even a slight sensation of disturbing thoughts or emotions serves as a motivator for scheduling a time-out period for reflective un-busy-ness. Allow yourself the luxury of not having to constantly do. Flavor your life with the sweet harmonics of Oneness.

Eventually, experiential Oneness will be embraced by all humans. However, until they reach that level of cosmic maturity, individually and collectively they will continue to inflict onto others what is perceived as personally unfair or burdensome. Eventually, they will learn the futility of their ways—although it has taken thousands of years to do so—and there will be no more overtaxing of one to serve the interests of another. Understand: fatigue and anxiety are measures of instability unknown to beings who maneuver chariots of light.

We suggest you routinely run self-monitoring assessments. Clearly ascertain where levels of even minute instability exist. There is an art to equalized energy distribution—a cosmic principle of perceptual balance practiced by finely tuned dimensional beings. As always, routine meditation is essential for working with and understanding fluctuating levels of solar and cosmic energies.

Most humans are not gifted at inward tuning or at adjusting to influxes of unseen energies. They are only slightly aware of how the sun and the cyclical nature of the moon affect their daily lives. Although they do somewhat recognize the power of the sun, they are generally unaware that the sun has the capability to enlighten or that the sun is a Divinely established receptacle of Universal Wisdom, a true cosmic library. They misconstrue the sun's underlying role within an inhabited solar system, relegating it to its most obvious but not necessarily most important function—to nurture and promote life. The sun's ability to sprout life from a seed is a symbol of its ability to give birth to Universal Oneness, or Cosmic Enlightenment. At this time, solar energies are increasing in relationship to increased energies spreading throughout the universe from the region of Central Sun. Generally, humans resist the influx of these energies. Most have not learned to flow with and absorb them for the greater benefit of all beings. Eventually, all will learn to do so.

The ability to secure and anchor light into the planetary grids is a preliminary exercise for adapting yourself to the influx of refined energies. You who are activated, spiritual light-workers have accomplished much in terms of healing and reconnecting Earth grids, but we caution you to be watchful for a downward drag that can be associated with these energies. When you are feeling anxious or fatigued, we suggest you back off for hours or even days, take time off, and indulge in playful recreation with close attention to diet, exercise, and sleep. Meditate silently in a place of absolute stillness. Practice healthful regard for the sacred temple that is your life's physical vehicle.

Transitional times are particularly karmic-sensitive. Therefore, routinely cleanse the contours of your energy bodies. Go beyond astral body and think emotional, causal, and mental bodies. You may feel drawn to expand even further, to Soul body. However, a certain degree of focused attention is necessary to perform this exercise. We suggest you keep it simple until you attain a level of proficient energy-body cleansing. A meditative chakra shower is very effective.

As the physical body requires routine maintenance to keep it running efficiently, so do the energy bodies, which are easily refreshened with a daily dose of light. Focus on a glassy, blue-white ball four to six feet above your head. Envision a shower of light pouring over your head, down your spine, and out to encompass all your energy bodies. Scan your energies outwardly to ascertain where manipulative fragments of karmic thought forms may have attached themselves to your bodies. Douse their outlines with a shower of light from the blue-white ball. Saturate every pore, every cell of your physical body with light. Expand outwardly until every aspect of your energy bodies is encompassed in this cleansing, healing light. As thought creates form instantaneously, it takes but a moment of inwardly focused intention for routine energy-body maintenance.

The tendency of sleeping humans is to project most of their energy outwardly. This is one of the primary reasons they have remained cosmically immature. Because they are simply not in the habit of deep levels of inward reflection, their brain-minds are basically disconnected from their heart- or Soul-minds. Extensive spiritual lassitude permeates contemporary society. This has resulted in epidemic levels of Soul hunger, particularly in technology-oriented countries. Emphasis on training the logic centers of the left brain has closed down right-brain cosmic portal access in all but the spiritually creative. Note the emphasis on spiritual creativity—that which is manifested in Oneness when intent on a cooperative state of higher-world connection. Humans tap into upper-dimensional contours to converse with light-beings primarily through forms of mystical art. We warn you, that which you presume to be rational is not necessarily so. The level that humans can reach to access higher knowledge is directly related to their spiritual maturity and ability to absorb and maintain refined cosmic energies.

Higher-dimensional apparatus to assist you who are readying for ascension is in place. Light-beings welcome those who prepare for that moment of Divine triumph when they assume light-robe attire, identifying them throughout the cosmos as awesome beings who have broken through chains of darkness to set their Souls free.

The Importance of Self-Discipleship

The natural glow of our energy bodies brightens considerably when we make telepathic contact with humans who are learning to detect the silent whispers of their inner voices. We are pleased to assure you that legions of starships have transported beings of light from many star regions to assist you in your struggles to evolve. It is our delight to cherish and comfort you at all times.

You may justifiably inquire why we are so solicitous of you. Why? Because you are One with that which are we! You are members of our beloved cosmic family. Like mighty eagles, you who are about to take flight into the vast spatial skies will soon land upon worlds that know no encumbrance to thought or greatness of Being.

We have emphasized many times that to push outward and upward will require a great deal of effort and self-discipline. Though we stand ready to assist you, it is your task to transmute your physical mass to light. As do all endeavors of great worth, you must understand that a large amount of determination, perseverance, and enthusiastic commitment will be required to achieve your goal. You must stay alert to outmoded habit patterns and repetitive ways of reacting that tend to pull you backward.

You have taken a holy vow of self-discipleship. Essentially, as a self-appointed follower of Divine Love and Light, you have taken a high-magnitude pledge. When you urged yourself to awaken, you took a step toward unlocking the windows to your Soul. For a third-dimensional being, this is the most profound journey you could embark upon.

As your extrastellar contacts are urging you, an earnest willingness to meditate, intention to activate the intuitive mind, and respect for every facet of your being are critical. You have committed yourself to Self-realization with a goal toward God-realization, Christ Consciousness. Because That Which Is Everywhere has always resided within you, by definition you are already a vessel of Divine Omnipresent Love and Light. Focusing on your expanding awareness of such things, learn to be compassionate and loving toward all beings and toward yourself as well.

Humans have reached a point of unrealized potential and are being challenged by the galaxy to make a conscious commitment to vibrational refinement. Like one massive butterfly, you are being asked to shuck your chrysalis of cosmic ignorance.

As you incorporate the teachings that weave in and out of the Arcturian Star Chronicles (manuals designed as galactic instruction packets), take note of their durability. They contain basic ingredients for living a truly good life.

They are structured upon spiritual texts natural to all human cultures. Sacred writings and teachings are always a cooperative effort between avatars and universal brotherhoods of light as orchestrated and transmitted from Absolute Boundless Being.

These manuals are purposefully repetitive. What is written here is restated in many books and in many spiritual workshops. Galactically inspired messages are issuing from the mouths of babes and from those of innocent bearing. *This is the most critical age in human history.* It is imperative that we catch your attention long enough for you to begin to assimilate the urgent nature of the times you live in.

We are well aware that humans are resistant to outside interference in their affairs. It is also evident that concepts such as self-discipline and self-responsibility often strike discordant notes within the mass psyche. In spite of these difficulties, we will persist in using as many methods as possible to alert you to the magnificence of the times and to remind you that you are an essential co-creative feature not only in your own evolution to light but in Earth Mother's as well.

Light Grids

Divine Aum tones and melodious aromatic scents are foundational ingredients for manifesting light grids. The stellar light grids—a star-to-star linking network—are saturated with honeyed vibrations as delicately hued as butterfly wings. Subtle odors that swirl and blend with the songs crystals sing are essential, for combinations of essence of Aum and space rose create a slight rippling effect of sparkling, many-colored lights that dance along the entire length of the stellar grids.

The grid coordinates are carefully orchestrated blends of color, light, sound, and aroma that move throughout the system with a wavelike motion which keeps the grids glistening. Clasping stars, planets, moons, comets, asteroids, and cosmic dust in their ethereal embrace, overlapping grids intersect with and bypass all lower-spatial time restrictions. Grids are also omnipresent webs of crystal-clear light that weave through dimensional transference points. They hold galaxies and universes in perpetual Oneness with Impeccable Cause. They vibrate and hum in joyful exaltation of energized Love, the music of the spheres, and Divine Intelligence manifesting as Light and Sound.

Aum-quality sound can be executed at the third-dimensional level by gently striking the edge of a crystal bowl. Crystal bowls, crystal wands, crystal balls, crystal clusters, and crystal points absorb waves of cosmic light that the crystals then transform into energized bits of information. These data are then stored in the crystal's sound reservoir until accessed.

Grid conduits will be explained in detail in a later chapter. In this essay it is sufficient to prepare one more step up in the knowledge of universal structure and how foundational matter comprises coordinates of color, light, sound, and aroma. This information is necessary if humanity truly desires to make technological breakthroughs and acquire knowledge of and privilege to interdimensional travel via star portals. The interactive dynamics of sound and light must be understood from a spiritual level before humanity is allowed multilevel entry by the Regency star council.

The notion of individualism that keeps human consciousness stuck in third-dimensional science must give way to a broader acceptance of the universe as a living entity. This must be done through a devotional approach to color, light, sound, and aroma as innate properties of Primary Energy as set forth in ancient religious texts. Space travel must be looked upon as a going forward to God, not as a going forward to explore what is already known to the vast majority of this galaxy's inhabitants.

Do not fall back in dismay, but look forward in joy to That Which Calls you. We of the great brotherhoods of light are reaching down from dimensions of blissful peace to issue you an invitation to the stars. What you must ask yourselves is, Am I ready to accept this invitation and all it will require of me? Those who are not will not receive a similar summons for a very long time. Individuals who have chosen the stars, humans who are advancing spiritually will, ere long, board finely vibrating spacecraft from many star systems. They will be granted unlimited opportunity to explore planets and stars of untold systems.

Is it your intention to graduate Earth school and merge in conscious Oneness with beings of light who travel freely about the space grids? The choice is yours.

Adonai, Sweet Ones.

The Essence of Being

On the eighth day He came into Being. On the ninth day He rested. His purpose was to transpose Sublime Memory into melodious octaves of magnificent light and sound. His journey is reflected in those of the star children whom He created at the beginning of the beginning. As above, so below. What Supreme Omnipotent Omnipresence set in motion is eventually experienced by all elements of Creation. The common factor of that experience is Oneness of Being. Any sense of separation from One is the result of a foggy substance, best described as illusionary mist, that shrouds the spiritual eyes of beings who exist at the lower depths of the cosmos.

Human, you have felt you were lost, struggling as you do against the relentless tides of forgotten Soul Memory. But never were you so. Do you not recall? You are Soul, an eternal pearl of unblemished splendor. Your Self knowingly resides in perpetual Grace. As a third-dimensional incarnate being, you are Soul inhabiting a transitory body for the purpose of facilitating an enhancement of Light Essence, a common denominator of Souls choosing to incarnate for a period of universal time in lower-strata galaxies.

Because a multitude of star-essence beings established residence on Earth aeons ago, a certain level of negative/positive harmonic balance is sustained by their diverse energies. This situation is not altogether unique for a lower-domain world, although humans like to think of themselves as particularly special—as indeed they are. The quality that ensures Earth's placement within the corridors of spatial time and that distinguishes it from other third-dimensional planets is the multiplicity of starseeded Souls who live there. Cut off from the vast cosmic playground that is their natural habitat, Earth's inhabitants have been kept secluded from their home worlds for thousands— in some cases millions—of years. Because Earth's evolving population is readying itself to assume aware galactic citizenship, the great brotherhoods of light are directly available to assist them.

To encourage a renewed sense of ecstatic hope in humans, from time to time That Which Casts Out No Thing incarnates a microcosmic portion of Itself in the guise of an advanced spiritual master, or avatar supreme. This is to ensure that Its cosmically unconscious portions— such as humans—have access to information that will assist them in uncovering their Soul Memories and provide them with a detailed map to Celestial Home.

Earth inhabitants and lower-astral entities have free will and may choose to embrace positive or negative attitudes or a combination thereof. No matter; Soul never loses awareness of itself as a sparkling bit of sustained Source Light. The concept of sin as preached by the fundamentalist religious traditions is a ploy the Dark Lords use to entice humans to accept such mischievous philosophy. They satisfy their malevolent purposes by absorbing pulsations of negativity emitted by humanity's dark side, the negative ego. Their capacity to reduce human light emissions lures the unwary into their dungeonlike playgrounds in the lower-astral regions described in religious texts as purgatory, hell, and the lower bardo. In fact, "sin" is a lower-case acronym for "stuck in negativity."

After the great galactic war, when Eden-like Earth was invaded by dark extraterrestrials, the Spiritual Hierarchy placed a blanket of forgetfulness—a holographic energy belt—over Earth's astral regions. It was put in place to restrict the comings and goings of all manner of unholy beings. To counteract the resultant demise of spiritual energy surrounding Earth, angelic guardians maintained the upper-astral levels in a state of perpetual bliss, described in sacred literature as the heavenly realms or Nirvana.

To unlock the karmic chains that have kept Earth's inhabitants cycling between repetitive astral and physical incarnations, advancing humans are urged to constantly challenge themselves to reach a level of sustained vibrational refinement, essences of light common to upperworld beings. This is best accomplished by using meditative and creative visualization methods.

Adonai.

Blessed Are They...

80

And there came out of the night a Holy One, and He spoke to His followers thusly:

Blessed are they who mourn for Earth and the citizens thereof. Blessed are they who seek cosmic justice in the form of compassionate unconditional Love for all living things. Blessed are they who call upon the angels and the many brotherhoods of light to assist them. Blessed are they who program Christ Consciousness into their hearts and minds, for they have agreed to assist in Earth's transformation to light. These blessed beings are known as warriors for the light, eagles of the new dawn, and ere long they shall know peace and they will be comforted.

Blessed are they who call Divine Presence to them and rejoice in Its Light. Blessed are they who in spite of the cares and sorrows of their days adjourn each night to the warmth of their beds and sing praises for the gift

of life. Blessed are they who respond to others with loving understanding. Blessed are they who gently reach
out with open hearts to touch and nurture others.
Blessed are they who wear mourning's black cloth as
their beloved ones leave their cares behind, their work
in this life complete. Blessed are they who stand tall as
trees and rise above the forest of despair that darkens
the world's peace. Blessed are they who welcome the
morning with a song upon their lips and a joyous attitude. Blessed are they who live in trust that all will
eventually be well. Blessed are they who despair the
loss of Light. Blessed are they who feed hope and
encouragement where none can be found. Blessed are
they who persevere upon the path of justice and harmony. Blessed are they who count their pennies wisely.
Blessed are they who look to their families and friends
as teachers and guardians of sacred trust. Blessed are
they who find life's temptations not half as sweet as the
temptations of Spirit. Blessed are they who whisper
from their hearts' core that they willingly join the
galaxywide effort to bring peace and prosperity to Earth.
Blessed are they who take the initiative to become all
that they already are. Blessed are they who know that
Light's Essence resides within them.

We are One! We come in delightful readiness to spring
forth radiant joy. Come into the House of the Lord of
Oneness. Come! Embrace the vision only mystics heretofore dared dream. Do not speak with tongues of solemnity
as do prophets of doom, soothsayers, and wisdomless

beings. Intuit the future from a place of abundant hope, for your dreams and visions are replete with your awakening Memories of many suns and the hum of many planets.

Within the hearts of humans rest alchemic ingredients to sever themselves from the bondage of aeons. Look within. Within is the way to freedom and light-body liberation. These words are crucial for your understanding of the Oneness of All Things.

Adonai.

Elohim and Hybrid Elohim

After the fall of Atlantis and the reestablishment of humans on Earth, newly birthing humanoids welcomed and openly interacted with beings of light. Awed by the latter's dominion over the forces of nature, the people assumed that those beings were gods. Countries in the hot climates of the Middle East such as Palestine and Egypt, countries bordering the Mediterranean Sea, and others such as India were particularly conducive to the interaction of beings of light and humans.

Before humans understood there is but One God, they believed in a multitude of gods. They referred to them collectively as elohim and singularly as el. Spiritually unsophisticated, in their innocence humans erected stones and statues to their favorite gods and participated in rituals they felt were necessary when beseeching a god to bless crops, to protect families, to bring healthy children, or to win victory in war. Soon, the more powerful elohim became established in heroic mythologies as Thoth, Zeus,

Hercules, Thor, and others. Not all beings visiting Earth were of the light, however, and many demanded allegiance and sacrifices. These occurrences are well documented in humanity's sacred literature and early writings.

So as not to frighten people steeped in superstition, during their periodic visitations beings of light often assumed human form or would take on the appearance of an animal, a snake, a burning bush, a tree or rock. These encounters were highly charged for humans—often the supreme moment of their lives. It was only natural that these experiences were woven into sacred stories and were among the people's most cherished possessions.

The true history of early humans is embedded in their traditions and mythologies. Recorded there are events whose origins can be traced back farther than Lemuria and Atlantis, back to the beginning of the beginning when Earth was seeded with life from the stars, back to more splendid, glorious civilizations that existed on Earth before the galactic war trapped humans and erased the Memories of their starry homes. Though your logic-linear-minds may scoff at these pronouncements, your heart-minds may very well find them hauntingly familiar.

In current-time human affairs, elohim serve as galactic emissaries to Earth. They are members of Earth's Spiritual Hierarchy. They are creative directors on the brotherhoods of light councils under the auspices of the supreme One, Sananda, the Christ Essence.

Now, as beings from the stars confront you with the knowledge of your true origins, they are putting forth the cry to all humanity: "Cease to be lost! Cease to be lost!"

Because reincarnating starseeded humans were trapped and could not maintain themselves in cosmic Oneness, they have always struggled with the fragmented energies of duality. Starseeded dolphins and whales are somewhat protected by their watery homes. The cloud of unknowing does not penetrate water as readily as it does air. Subconsciously, this is one reason humans are so drawn to water and are comforted by bathing in it or gazing into its depths.

Before Earth bore life, elohim came as creator gods to drop motes of light from great starships upon desert and mountain. Light also penetrated deep into the oceans' depths, "planting" microforms of cosmic energy destined to sprout into millions of plants and animals to cover Earth with a magnificent gown of verdant life. Under the watchful eye of the Glorious One, these elohim have always served in harmony with Universal Oneness to seed newly establishing planetary systems.

In the beginning years of the Earth-seeding experiment, all went well. Those who settled there knew who they were and from whence they came. Then a great darkness fell over Earth's inhabitants as a vicious interplanetary war in this galactic sector came to an end. As the Dark Lords gained control, they severed humanity's cosmic Memory connectors. Initially, humans could not grasp what had happened to them. One moment they were awake, and the next they were thrust into confusion. They were like amnesiacs. One moment they were One with the stars, the next they were completely incapable of pronouncing their own names. Thus, the long sleep began.

For many reasons, once in place the veil of darkness could not be lifted. Thousands of years would be needed for its energy to dissipate. The desolation caused by the placement of the veil was similar to the time it takes Earth to heal herself from a dose of intense radiation poisoning.

Those who remained aboard starships outside the Dark Lords' range did not fall prey to the Memory loss. Throughout spatial time they have remained, watching over, protecting, encouraging, and guiding those who were slumbering.

Over time, many Souls have come to Earth that were not part of the original starseeding. Some came intact, only to succumb to the darkness of forgetting. Advanced beings, however, have learned to retain Soul-knowledge integrity. They are able to penetrate the darkness, remain on Earth, and return without harm. Early humans had the advantage of recognizing those golden beings. They saw them as gods who came and went from fiery chariots that resembled the motes of a thousand suns.

Suppress a smile at the temerity of our scribe to put words she barely understands to paper. Dubious as they may be, they are, nevertheless, encoded with thought forms of cosmic wisdom. Today's human can barely grasp the powerful contours of beings of light long ignored by a people who hold no tolerance for such things.

Awakening humans are rediscovering the timeless presence of the elohim. These humans are intent on reestablishing the connecting link to their stars of origin. Intuitively, they know that this is the most auspicious life they have ever led in human form. Eager to activate Soul

86

SANAT KUMARA, MASTER OF LIGHT

purpose, they are pressing themselves into service to the Omnipotent One. Elohim they are becoming, ascension their goal.

These vibrationally refining humans are hybrid elohim, starseeded beings whose awakening hearts have no tolerance for a compromised mission. (Hybrid elohim are not to be confused with the hybrid race being created by the Zeta Reticulum.) They are determined to reopen their spiritual eyes. As they learn that no displacement or demise of light is possible under Universal Law, they are losing their fear of death. They know the Law of One is unchanging and remains as exact as when Moses came down from the mountain clutching sacred tablets on which were written the tenets of the Law.

From time to time starseeds catch glimpses of when they bedded down in tents that smelled of camel dung. They remember when they walked the lands of prehistoric America. They are beginning to realize they are hybrid elohim, and they have petitioned beings from the stars in the silence of their dreams.

Do not compromise yourselves with preconceived notions of godlike beings who freely interacted with early humans. Read between the lines until it dawns on you that you sit with your opening eagle eyes focused on the natal cord of a planetary rebirth. Your budding intuition is your link to the elohim and to other hybrid humans. You are children of the stars. Like eagles in flight, you scan the world for others who, like you, are arousing from a long night of cosmic sleep. Among you are incarnating elders from Earth-honoring traditional people. You are the anointed ones. From your position of spiritual maturation, you can clearly see the Eye of the One. As you gain understanding of your Souls' origins—which eludes those who walk the trail of sleeping people—your purpose for being activates.

Move quickly beyond any anxious feelings induced by the ramifications of this essay's message. For many years we have warned you to expect the unexpected. We tell you, you are hybrid elohim becoming elohim—creator gods in your own right. Holy! Holy! Evolved galactic communities are entranced with the subdued light Earth's awakening elohim are emitting.

_____

Comes the One called Christ by Westerners, known to Easterners as Buddha, the Enlightened One. My I Amness at higher levels transcends the confines of human perception that limits the Holy of Holies to a personality associated with only one philosophy. Ab-Soul-luteness of Being transcends all limited belief systems. I-Amness-Buddha-Christ emanates Light that shines brighter than the rays of the most brilliant sun. I am known as the Enlightened One. My ancient teachings persevere as written and oral sacred scriptures. Throughout time I have cloaked my I Amness in many guises. Nevertheless, that which was given to early humans remains of paramount importance.

I Am That I Am, radiant Christ Essence, the Dove, sometimes known as Lord Sananda, Keeper of the Immaculate Ray, the Flame of Life. My I Amness intercepts and interprets Divine Light for developing star systems. I Am that which covers Earth with a cloak of brilliant Light. I Am Essence of One, sometimes known as the

Immaculate Brother of Heavenly Lights. I greet you from the Pleiades. I greet you from Earth. I greet you from Arcturus and distant Andromeda galaxy. My I Amness is contained in all things. That Which Is One is indeed One, now and forever. I have embodied upon Earth many times, in many forms, to establish the simple dynamics of unconditional Love as Divine Law on Earth.

Regency Star Council Transmission to the Awakening

Ladies and gentlemen of Earth-energy arena, you have all been brought into this sanctuary, this Garden of Eden, to take part in a glorious celestial experiment. You who peruse the pages of this multilevel telepathic document have already been introduced to extrasolar beings of light. Although you may have long forgotten your connections to the stars, the cells of your bodies quiver with the stirrings of awakening. Spiritually flaccid from lifetimes of disconnection, your subconscious attunement to the many suns of this glorious galaxy is dissipating the clouds of forgetting that, like heavy curtains, have long covered your Soul Memories. Like butterflies emerging from cocoons, you are unfurling and testing your magnificent spiritual wings. You are being reborn into awareness.

Implicitly trust your Self to lead you on a pathway of optimum conditions that will enhance your progression to light-body status. As a seeker of Light, you must diligently and persistently strive to achieve this most worthy goal. The massive energy adjustments that are urging Earth's body to a higher vibration are affecting you as well. The opportunities she is receiving to evolve are equally yours. Your most immediate concern, however, is to strengthen your compassionate stance and farseeing objectivity in the face of escalating reports of the doom-and-gloom Earth changes. Your ability to alchemically transform is your intuitive awareness of the need to maintain yourself at specific levels of vibrational resonation at all times. Because of this, you are urged to quickly assess your personal challenges and to move through them as soon as they confront you.

Vacation is over. No immediate relief is seen. Buckle down and tackle the job at hand, a position for which you long ago volunteered. Mechanisms to trigger your awakening have already been activated. If this were not so, the scope of these transmissions would be beyond your comprehension. As an awakening starseed in service to the evolution of Earth, you are among the cream of the crop. You are an unqualified expert in your field of galactic expertise.

The situation is primed to test your integrity and ability to sanction high purpose within lower levels of gathering darkness and to lovingly and intentionally focus Light. The cancerous condition overtaking the spiritually inept is like a plague that has caught the unwary off guard. You are witness to many expanding timelines that include an eruption of negativity ensnaring many humans in the Dark Lords' growing web of deceit.

This document is sealed by the Regency star council, Sananda presiding, and may be considered a personal directive. Our hope is that its contents will enable you to free yourself from any remaining difficulties that tend to pull you backward. New timelines are being set in place. Knowledge of how to use these expanding cosmic energies will greatly assist your journey upward into the domains of Light.

Keep your intuitive channels open so your personal galactic assistants can easily contact you regarding even the most mundane matters. The details of your daily life contain enhancements for your progression to the stars. As One with other like-minded individuals, collectively you carry a powerful torch of transformative change. Your efforts are building blocks for a spiritually and cosmically transformed human society. Your children and your children's children will be the benefactors of your good deeds. Surely it is a very fine thing to be a pioneer of a new-dawn world.

It may or may not surprise you that your companions on intergalactic starships are depending on you to expedite your assigned tasks with gusto and good cheer. Look upon what you have been asked to do by that nagging inner voice, which constantly urges you forward, as God inspired. As we note the results of your good works, we observe beacons of effervescent light, as spectacular a display of fireworks as has ever brightened a holy-day night.

A special-forces intergalactic team empowered by Earth's Spiritual Hierarchy under the auspices of the Regency star council has granted permission to your Soul counterparts from the upper-universal regions to directly assist you. It is assumed you have already made contact, or are making efforts to do so, with your higher Self. It is also assumed that you periodically access external sources, such as like-minded friends, books, movies, and the Internet, to confirm and perhaps update information directly received.

Adonai.

Sweet Transformation

Elevators of light are rising from layers of karmic debris that aeons of multiple lives have settled upon you. Sweet transformation! Urge yourself above the stresses and strains of your life and think of yourself as One with all Creation's perfection.

The time for Earth Mother's ascension into refined light is near. Humans who maintain themselves at or higher than her evolutionary vibratory rate intuitively feel the great sweeps of cosmic energy that, for some years now, have been inundating this sector of the galaxy with high-magnitude light. Humans who are preparing to ascend are like torches of fiery light shining brightly for others to follow. Dedicated spiritual practitioners are assisting star-masters in transposing Earth's physical components to effulgent light.

By now the adept have undoubtedly been challenged by unsavory entities who attempt to deflect them from their path by instilling them with fear. In a last-ditch effort to halt any forward progress, they flatter, distract, and coerce the spiritually determined to give it up. Pay them no heed! They have nothing of value to tempt you with! Their feeble ploys cannot satisfy the hunger of those who seek God.

Your stated commitment to spiritually evolve not only is physically transformative but affects your DNA and your energy bodies as well. Your body is attempting to simultaneously tense and relax to release itself from the restraints of lower-plane manifestation. Your powerful lower self is naturally resistant. Ego's motivation is to relieve itself of the discomforting sensation of dwindling lower-tonal energy as the powerful higher Self takes precedence.

Your physical body is struggling to absorb an onslaught of cosmic energies it has never before been subject to. As a result, it is apt to become afflicted with unusual flulike illnesses, muscle and joint pains, skin rashes, newly acquired allergic reactions, and so forth. The marketplace proliferates with alternative and traditional healing techniques, books, and products. Recuperative efforts are enhanced and even accelerated with toning, meditative affirmations and mantras, periodic full-body massages, creative dance, tai chi, hatha yoga, herbal therapy, methodical energy work, dream yoga, and awareness of your body's reaction to all ingested substances.

Your self-motivated generation is more aware than any other of the profound effects that mental and emotional attitudes have on the well-being of the physical vehicle. As spiritual practices become a substantial part of your life, dive deeply inward. If you feel drawn to do so, consciously bond your energy with the star or stars to which you feel a yearning connection, an indicator of spiritual Home-sickness. Like strings tied to a puppet, starseeded Souls were attached to their stars of origin prior to venturing Earthward. Discerning your extrasolar connections will help you integrate your etheric DNA with your physical DNA. Correspondingly, your ability to stay comfortably centered and balanced will be enhanced by even a tiny bit of knowledge of your journey through the uni-versal (one song) harmonics.

Souls migrated to this solar system and took up resi-dency on Earth for many reasons. Nevertheless, they became captured in the interlocking energies of a force-field that long ago was set in place around Earth's astral regions. That restrictive veil is now dissipating; karmic cords that held Souls within Earth's energy fields are turn-ing to powder.

Beloved ones, you are beams of golden light. Your bril-liant auras radiate arcs of multihued rainbows. Collectively, you carpet Earth like a field of spring flowers. Yet, you who are about to access the stars continue to torment yourselves with self-other criticism, fear, and judgment. You who indulge in one transitory adventure and tasteless morsel after another are beings of suppressed light. You are indi-vidualized notes in a magnificent celestial sonata.

Extrasolar entities do not look upon you as ignorant barbarians. We know how you struggle. We are aware of your difficulties. But you are seedlings ripe for bursting. Delight in what we tell you. We are outlining the contours of your true nature—Soul.

In meditation, focus on absorbing Christ Consciousness as a third strand of the double-helix DNA. Note that the third strand resembles a ladder. Without hesitation, climb up the intertwining spiraling columns of the multistranded etheric DNA. Now observe yourself boarding a crystalline starship where you are heartily greeted by multistellar beings of light.

As you go about your day's busy-ness, the reality of your starship visits will take on dreamy, otherworldly qualities. In spite of this, you have made conscious contact with entities who live in realms of perpetual enchantment. Those who come forward to introduce themselves have taken a holy vow to encourage and assist you on your spiritual journey.

Addendum: There is an unfortunate tendency among awakening starseeds to restrict members of the great brotherhoods of light to various titles and nonexpansive levels of being. Relax your tendency to categorize angels, ascended masters, and multidimensional extraterrestrials. Holy personages do not consider themselves as separate. In every sense we are One. You are One with masters of light and one another as well. You will not reach the level of Christ Consciousness enlightenment through separative behavior. Creator is One integrated field of Intelligent Light. All things are One and all things have One purpose—to merge with Omnipotent Source. Although Energy of One retains individual consciousnesses, It is in agreement at all times with unity of Universal Mind and maintains Itself in ab-Soul-lute knowledge that all things are emanated projections of Omniscient Divine Light.

We are aware that our teachings and our manner of expressing ourselves can be difficult to understand. We state this as simply as possible: You are an aspect of Omnipresent Universal Consciousness.

Awakening Masters, Come Forth

Sorrow spent, Earth's ascension into realms of harmonic tranquility is assured. You who intuitively comprehend the truth of this statement are among Earth's oldest Souls, hybrid elohim, members of the Holy Seraphim inhabiting human bodies in the dream state. Your purpose is to cooperate with your associates within the great brotherhoods of light who "use" your physical vehicles for impregnating Earth with beams of transformative light. On the surface, it may seem your free will is being bypassed. However, we assure you it is not, for our mutual interaction is in compliance with a contractual vow your Soul made prior to incarnating onto Earth.

Transparent as lace, the etheric bodies of those slated for ascension are exquisitely bright. Only a few shadow remnants of lower-ego energies remain in the spiritually alert. Nevertheless, many still have their work cut out for them, as the details of the gentleness of spirit indicative of a Christ Conscious individual have not yet completely manifested in third-dimensional reality. But in time that, too, will come.

You who feel an urgent summons to fulfill higher purpose are awakening masters. Your heart chakras are opening with genuine compassion and unconditional Love. You are wiser and more capable of spiritual discernment. You have grown in your ability to practice Universal Law. You have achieved a higher level of Self-realization. Still you press on, never hesitating to complete your goal of assimilating light and consciously operating as an absolute God-aware Being.

Humanity is moving closer to an inevitable moment when it will be forced to accept or reject its invitation to become a conscious citizen of the greater galactic community. One task of the awakening is to help prepare the masses for that day: to inform them that they are not in any real danger of invasion by monster-shaped extraterrestrials; to awaken them to an awareness of multidimensional reality and the mandatory aspects of Universal Law, which are joyfully and willingly practiced by all citizens of spiritually advanced planets.

In 1998, Earth set her course along a variety of alternative timelines (see in Part IV, "Grid Conduits, Emerging Timelines, and Alternative Futures"). Those who have prepared themselves for Armageddon will certainly experience that horrifying scenario. Those with a higher opinion of themselves, that is, as Soul having a human incarnation, will never again succumb to fear's unsavory ways—unless, of course, they renew their willingness to volunteer for developing-planet duty. Never again will they question the elementary fact that they are highly cherished universal beings. They will understand that there is no higher-world port of call that prescribes lower-state energies as necessary evils.

The resonant covering that lies over Earth's lower levels is clearly documented in records kept secret from the bulk of humanity. Only a few of Earth's behind-the-scenes historic leaders have been privy to this information. They neglect to inform you that you have been kept subject to a species of entity who has deliberately separated you from the greater galactic community. Aware planetary citizens make up the majority of galactically inhabitable regions.

The energy belt that was set in place around Earth at the time of the fall of Atlantis is the primary reason your radios and telescopes have not been able to make outside contact. Light-frequency energy where the majority of the universe's citizenry keep themselves tuned is far too subtle for your current technology to pick up. Your scientists resist understanding that crystalline energy combined with music's tonal qualities and your own latent telepathic abilities are primary components for accessing spatial-time travel. The brotherhoods of light are attempting to demonstrate critical transformative scientific information via crop-circle formations. Many crop circles are technical replicas of instruments used by galactic scientists. Among other things, they record mathematical equations from a cosmic standpoint as the vibratory resonation of stars and planets—the music of the spheres.

Humans who appreciate themselves as being outside the unyielding dictates of linear time—that is, awakening masters—are delighted with the conditions of Universal Law and the release of an abundance of useful galactic-level information that historically has been kept out of human hands. The aware are learning to move beyond third-dimensional spatial-time restrictions. They are learning to incorporate light into Earth at an increasing megahertz rate. Changes in Earth's vibrational status could easily be verified as escalating within light's known spectrums. In fact, her light-body has begun to resemble the bursting flares of solar gas that periodically erupt from the sun.

It pleases Earth's spiritual management team to make data easily available to you to assist in your progress toward spiritual enlightenment and your awakening as a citizen of the universe. We wish to increase the knowledge of your incarnated status as a member of a multistellar galactic task force. This information is being given as a means to enhance your spiritual self-empowerment. It is not to be used for self-aggrandizement or ego enhancement.

Earth's Destiny

Our theme is of Earth's transition to light and humanity's attainment of the stars. You are hereby challenged to kindle and activate high purpose. Masters of light are confronting you with the necessity to transform your separative behavior into applied universal Oneness. You must integrate your societies into planetwide wholeness before the spatial-time disparities that cause you so much difficulty are completely withdrawn. You must understand that Earth is a Soul-challenging, teaching planet. That which is trying to the human is of great benefit to Soul. If Soul fails to accomplish what it came to do, Soul will simply be sent elsewhere. This is similar to being sent back a grade or two in school. It can amount to thousands, even millions of years of time as humans know it. Time to Soul, however, is infinite.

Patricia's interdimensional telepathic recording phase is almost complete. The final manuscript of her four-part series, Arcturian Star Chronicles, has been co-created by

Patricia and several high-intensity beings of light. In fact, the final chapters comprise a series of direct transmissions by titanic star-masters from the depths of the Great Central Sun. Presently, these awesome beings are inundating this galaxy with wave after wave of extremely refined light and sound, massive inflows of cosmic energy affecting all who live on Earth.

The blending of ancient mystical knowledge and modern science, which is becoming so popular, has created a desire in the galactically aware to reestablish Earth as a commonwealth of the great star nation. It is this generation's birthright to hark to God's urgent summons and to valiantly commit to serve a cause whose primary dynamics lie far beyond Earth. In preparation for an announcement by the world media to be given early in the new millennium (no news organization to have exclusive rights), it will be established that what has long been predicted by those who possess clear vision—that is, the beginnings of a thousand years of peace—has become fact. The stability of this new society will be observable.

This announcement will be the final broadcast by current media, whose purpose is to confuse and instill fear. Thereafter, an era of joyful, perfect harmony will bring into play a true utopian society on Earth.

We have attempted to clearly state that individuals and collective humanity have several futures from which to choose. This is extremely important information and will directly affect your Soul's progress for a very long period of galactic time.

Those who have made the exquisite choice to prepare for ascension are scheduled to do so in conjunction with Earth's transition to light. Low-toned entities, however, who broadcast negative energy through the ethers will find themselves descending into what is best described as the hellish regions. This is not the result of Divine condemnation. It is a precautionary measure the star-masters use to protect newly forming light-planets from intrusion by beings of low-wattage energy. Left to themselves, the latter would attempt to penetrate the energy curtain that keeps delicate transitional Souls safe from these dark elements.

Many believe conditions imposed on the cosmically ignorant are unnecessarily harsh. Nevertheless, it is the star-masters' concern that nonevolving elements be distracted from surging upward whenever a massive evolutionary shift up the light-spectrum scale occurs. Earth and her inhabitants are not the only third-dimensional entities preparing for light-enhancement. Therefore, it is logical that certain standards be maintained as universal constants.

This is not new information. It may very well annoy those who are not enlightened as to the exactitude of celestial cause-and-effect measurement. As is true in any honest accounting system, there are no variables allowing for lax attitudes or sloppy bookkeeping in a Soul's Akashic Records.

It is not too late to climb aboard the celestial star-ships that are assisting Earth in her preparations for liftoff. Until final thrust occurs there will be a narrow margin of opportunity to catch a ride on the upward-surging energies. For centuries humans have been cautioned to prepare themselves for this eventuality. Now little time remains to complete preparations for this grandest of all cosmic events.

On Group Energy

Focus on Earth's star, Sol, as a ball of vibrant, fiery energy. Quickly move far into space, readjusting your vision until the sun becomes nothing more than a speck of light amongst countless other stars. From this broad perspective, contemplate Sol's placement in the greater cosmic picture. However unimportant Sol may now appear, if it were to suddenly disintegrate it would dramatically affect the stability of the stars closest to this system. Immediately upon Sol's untimely eruption, a tsunami-like wave of unstable energy would pass through the entire Milky Way galaxy. Thus, the integrity of the whole is predicated on the ability of even a minute particle of stardust to sustain itself as light steadily shining.

This pictorial analogy will help you come to a better understanding of how the whole is dependent on all its parts and how the parts are dependent on the whole. Simply put, the integrity of the entire universe is contingent on the self-sustainable energy of its many components: every galaxy, every star, every planet, and all the inhabitants thereof. Every little dot of light in the night sky is an integrated portion of a greater cosmos, a cell within the omnipresent body of Absolute Being. As above, so below.

Inhabitants of the multilevel universe resemble points of starlight in that they are formed from Soul substance. Souls are like stars in that they are mirror images of one another. Souls are replicas of Divine Light. As such, each Soul is responsible for sustaining itself in harmonic alliance with All That Is. Every Soul is a segment of Greater Whole, God expressing Itself as God, if you will.

May this visual tool bring you to a better comprehension of the importance of working with group energy. The integrity of any group is predicated on each participant's moment-to-moment ability to shine like a brilliant star. Regardless of the purpose of any group you attend, be willing to interact from a place of focused self-awareness. Be still when others speak and give them your full concentration. Pay close attention to intuitive timing, and when your turn comes, speak clearly from a position of consideration and appreciation for self and others. When all members contribute from their storehouses of diverse experiences, unique perceptions, and knowledge gained, they enrich, nurture, and sustain the dynamics of the entire group.

Unstable stars that refuse to share light and that instead pull inward are dying stars. Collapsed stars are dark, volatile, and extremely dangerous to their neighbors. Imploding energy can threaten the stability of an entire galaxy. On a much smaller scale, but not necessarily of lesser importance, is this: When individuals in group work pull their energy inward because of fear of appearing foolish or of being rejected, from an inordinate sense of personal privacy, or for any other reason, energy that is meant to uplift and sustain the group as a whole is dramatically lessened.

Nowhere are we-you separate. An individual who attempts to maintain a position of aloofness from the universal whole will never come to a true appreciation of life's purpose or a true understanding of personal em-powerment. An unimpeded flow of life-sustaining cosmic energy is more readily available to those who have the courage to be vulnerable and open to others. It must be clearly understood! Although the All is One and cannot lose It-Self in any fashion, it grants free will to all Its many parts. Therefore, when group members bond, each individual is able to draw larger amounts of abundant energy from the cosmic reservoir. This is evident to those who regularly participate in group activities.

Blessings on you who grasp the ramifications of this analogy comparing human groups to star clusters. This essay was derived from depositories of Earth's sacred knowledge. Libraries of universal wisdom are protected by beings of light of the highest order. On the surface, this focus on group energy may seem rather simplistic, but what is not as obvious is that humans must come to a clear understanding of the workings of group dynamics if they wish to interact in Oneness with higher-world beings.

During these dark years that are accompanying Earth's evolutionary transition, you must be able to support yourself in every fiber of your being with the absolute knowledge that you are Soul and that Soul is interlinked at all times with the Mind of Universal Intelligence. Soul, being unrestricted by the limitation of third-dimensional spatial time, is unequivocally in eternal Oneness with All That Is.

Awaken to a Greater Reality

Awaken to the knowledge that as a component of universal Soul you are a citizen of a great galactic community. Awaken to the knowledge that Soul is unlimited in opportunity to experience diversity of form. Awaken to the inherent possibilities contained in this statement. Awaken and remember who you truly are and get a sense of why your Soul chose to incarnate in human form during these auspicious times.

We who guide the affairs of the greater galactic system are filled with awe as we observe the awakening heart-minds of humans who have consciously pledged themselves to integrate wholeness of Spirit, to attain light-body status, and to complete greater purpose in this lifetime. Star seekers, you are coming closer to a moment in universal time when you will access the portals of Celestial Home.

As you learn to reach out to one another and acknowledge that you are truly One, you will be on the verge of touching infinity. Eventually you will come to the realization that infinity is nothing more than a prelude to the opening stanzas of an even greater reality. How can this be? You may trust us: It is so. As you awaken and reestablish your connection to Cosmic Intelligence, you will begin to process information from a higher level of awareness. As you do so, you will begin to identify with yourself as incarnated Soul. You will know beyond a doubt that the personality you now are is but a minute—yet vitally important—speck in relationship to the boundless being that is your higher Self.

Star seekers, we are pleased to observe your growing ability to properly access and use cosmic energy. As reflections of Eternal Light, you are awakening to the abundant, ever-present reality of Universal Source. Your etheric bodies have begun to shine as brightly as heavenly stars shine.

These are the end times of human torment, when sorrow stills and an abundance of peace and joy permeates through every region of Earth Mother's eager body. Appropriate to your level of awakening and your newly discovered ability to absorb Supreme Light into every cell and DNA strand of your physical body, your karmic debt as recorded on your Soul's Akashic Records is adjusting itself to zero degree—which is necessary to offset the cause-and-effect issues that have kept you recycling through physical and astral realms for untold ages. This is a massive step forward in Soul progress. You are justified in congratulating yourself on your accomplishments!

As a cosmically aware citizen, you will no longer be subject to the restraints of third-dimensional reality. The twenty-first century will progress while you carry out your daily routine with a higher degree of spiritual response, acknowledging Earth's entry into Light-level cosmic domains.

As you fully awaken to your status as a starseeded Soul, you will more actively participate in our efforts to reestablish universal order not only on Earth but throughout this solar system. For the most part, you will continue to observe your daily reality through your physical body's somewhat imperfect sensory systems and your personality's established behavior. However, your spiritual accomplishments have already begun altering the contours of your etheric body's energy patterns. As Earth frees herself from her heavy karmic load, eventually you, too, will completely release the restrictive bonds that for a very long time have kept you chained to a densely vibrating planet. We welcome the moment when Grace's sweet promise descends upon you and enfolds you in the radiant arms of perpetual rapture.

Through your will you are opening the doorway of your closed heart onto the universe. The moment your higher chakras begin absorbing and transmitting cosmically, you will automatically anchor and begin radiating light. Learning to focus abundant amounts of unconditional Love onto the planetary and stellar grids testifies that you have begun to remember yourself as a higher-dimensional being.

Awakening humans consciously endeavor to overcome their genetically manipulated tendencies toward aggressive behavior, or, loosely stated, "He who gathers stones had best not throw them." Star seekers humbly accept and acknowledge their divine natures as Soul. They are aware that they have never really been separated from Spirit and the active doings of the greater universal community.

Revitalize and reflect peaceful radiance into the eyes and ears of all whom you meet. Greet one and all with an open mind, a compassionate heart, and a generous spiritual bearing. Clearly honor the star child that dwells within every living being.

Go Forth in Joy:
A Commitment to Spiritual Living

Revered ones, you who aspire to complete your spiritual journeys without further delay, you who long to keep close company with the star-masters, from this day forth live your life in joy. Complete your intention to shed your unshed tears and to relieve yourselves of the heavy burdens of sorrow and vain ambitions that have kept you karmically tied to Earth. Conquer any tendencies that find you succumbing to despairing attitudes associated with backward spiritual motion. Monitor your thoughts. Honestly assess yourselves and any aspects of your lives that have lost their appeal and deflect you from your spiritual path.

Your associates within the great brotherhoods of light maintain constant vigilance over situations that concern every starseed. However, they are not always successful in their untiring attempts to telepathically communicate with you. Their efforts to initiate direct linkage are often ignored at the receiving end because of your inability to open your heart in childlike joy to multidimensional beings. Additionally, the human ego often manipulates multilevel information to fit traditional, stagnant belief systems. As starseeds expand their narrow conceptual boundaries, establish telepathic contact with their inner guidance teams, and become aware that Spirit's concern is that their highest and best interests are always served, cooperative interaction at both ends becomes the primary motivation for continued contact. The proof for starseeds is the positive effects their multistar association has on their lives. As they realize that they have a direct tap to the cosmic hotline, their joy becomes apparent. Certainly it becomes easier for them to keep themselves always alert while monitoring their thoughts for incoming messages. Eventually, as old karmic patterns are released and spiritual integrity becomes a way of life, the starmasters are granted permission to interact with their human charges from a uni-versal cosmic standpoint of highest and best for all.

Those members of the human family who are firmly entrenched in corporal illusion are digging a deep wallow of self-made, self-perpetuating horrors. The level of spiritual immaturity and cosmic ignorance the masses demonstrate drags down their spiritual reserves. What affects one affects all: It is next to impossible for awakening starseeds to completely relieve themselves of karmic patterns ascribable to current-day social influences and cultural conditioning without persistent attention to detail.

Your stated commitment to personal evolution, however, instantaneously aligns you with one or more beings of light. Thereafter, your progression to ab-Soul-lute harmonious alignment with Spirit becomes your step-by-step adventure. Accept that your Soul's primary purpose for incarnating on Earth in human form during these times is to use the opportunities associated with the closing and reopening of cosmic timelines on an evolving planet. It behooves you, then, to take careful note of your intention to interact with Spirit and to assess the level of your willingness to work in joyful tandem with beings of light.

Your beliefs regarding parallel worlds and alternative realities are not important. However, it is in your best interests to review your life and honestly evaluate your behavior patterns. Overhaul any issue that tends to divert you from taking an unwavering approach to dedicated evolution. Every effort made toward this end, no matter how minute, will eventually be rewarded. Soul reaps untold benefits in accordance with your demonstrable intentions and efforts to mature cosmically. Therefore, go joyfully on your spiritual path in full knowledge that what you sow each day will eventually reap an abundant harvest.

That Which I Am calls to you through time's mists. I come from the future as well as from the past. Call Me to you. My I Amness enhances and dispenses Source Energy through the light of the many suns. My I Amness is the Holy of Holies, bearer of the omnipotent ray of Ultimate Being. That which humans refer to as the lesser gods is certainly not that which dominates the Oneness of Being that humans somewhat euphemistically refer to as God.

Source Name expressed in English lacks resonant energy. Indeed, there are few Earth languages that adequately release omnipotent qualities of Buddha-Christ Essence, Galactic Home, Source Energy, Isness, or Universal Intelligence, let alone any words that have the capability of describing all things as One Thing.

Come Home! Come Home!

———————————

In solitude, alone (all-One) in meditative silence, you may have a vision of the universe's loftier realms, a dreamlike view of alternative planetary systems whose placement within the cosmos is more vibrationally refined than Earth's system. Although the dynamics of limited human consciousness do not allow you to physically move beyond the veil that keeps you separated from those worlds, there is nothing stopping you from freely exploring them with your inward sight. Concentrated inward focus connects you with the telepathic whispers of beings of light who are the inhabitants of those domains.

At times, you may intuitively feel you are being urged to move beyond the dreams and goals associated with temporal living and to seek your place among higher-world star systems. You may feel an urgency to establish a foothold upon the long pathway you inwardly know will eventually bring you Home.

An aspect of awakening for many starseeds is an intense and unaccountable sensation of Homesickness, an insatiable desire to return to some dearly beloved place left so long ago that they can scarcely remember it. Many have been overtaken with a passionate urge to return Home, often so powerful that its energies linger at the edge of their every thought. Calculating that today they will move themselves one step closer to their goal, they can relate to Dorothy, the Scarecrow, the Tin Man, and the Cowardly Lion. They somehow know they have stepped onto a golden pathway in an alternative reality and have no other recourse but to follow it to its destination.

Humans who are expanding their perceptions are getting a sense of starships gathering above Earth. Eagerly they scan the night skies for some kind of movement in the sparkling lights. Many now think of the occupants as beloved members of their families. They have begun to understand that they are attached to the universe through a system of cordlike energies, that Soul bonds them in harmonic Oneness with all other Souls. Sometimes patiently, often impatiently, they await their advancement to light.

The awakening are beginning to realize that they have received a summons to partake in a great cosmic festival. We have often indicated that all humans have been issued an engraved invitation. Therefore, prepare yourselves for that awesome celebration—the day the celestial starships reveal themselves.

Starseeds are companions of fabled angel-like beings. They are respected associates of light-entities who have voyaged to Earth from a multitude of star systems. Many experienced their first Earth landing at a time long faded from human memory when a gigantic cooperative interstellar fleet first came to your planet and a massive multisystem experiment was initiated. At that time, all beings came in peace.

While humanity struggles with the reality of unidentified flying objects as observable phenomena, awakening starseeds are more aware of who the occupants are and why they are coming to Earth. They do not hesitate to open telepathically and interact with them. They understand that a fleet of crystalline, multisystem starships is directly involved in the healing and evolution of Earth and her willing occupants.

Although starseeds recognize Earth as their planetary home while in the embodied state, their heart-minds brush up against a persistent summons to prepare for the long journey Home. And where might Home be? It's in the heart of the Central Sun, where beings of magnificent splendor rest in glorious eternal Oneness—the end of Soul's efforts to gather exquisite knowledge and experience.

The celestial highway is well marked. Soul knows the way Home. Humanity's task is to relieve itself of the layers of karma that have delayed the upward momentum. Deep within the core of all awakening hearts is a solid knowing that it is time to complete and release karma and to diligently pursue light-body ascension.

So beautiful you are. In spite of evidence to the contrary, humans are a naturally benevolent, compassionate, gentle species. But they have loaded their vibrations with aeons of disharmonic anger, pain, and hopelessness. In

these times, however, this unfortunate situation has culminated in starseeds reestablishing basic compassion and
mercy for all who suffer.

As karmic layers peel away, humans discover an underlying reservoir of absolute Love, a realization that they do
have the power and the ability to project waves of unconditional Love outwardly in gossamer strands of golden light.
When awareness of ab-Soul-lute self-empowerment is
activated, they know they will never have fear again.

If you were to ask our definition of human beings, we
would answer from Soul Essence observation. We would
emphatically state that they are made up of energy fields
of unrealized potential, that they are as capable of experiencing and radiating Light, Love, and compassionate
wisdom as any cosmic citizen is. We see them as substance
of Soul, as light in tangible form.

Those who diligently pursue their spiritual pathways
eventually reach a point when their attachment to ego and
physical form begins to fade in importance. As Earth is
swallowed in waves of extended Grace and held close in
the arms of Perpetual Rhapsody, karmic restrictors that
captured and bound humans to her massive orb will let go.
Then her inhabitants will experience peace as a constant
companion—not a sometime visitor that departs almost as
soon as it arrives. New-dawn humans will practice what
they preach and will love and honor their neighbors as
tenderly as they do themselves.

Those who read the material presented in this series of
books do so because they are committed to completing the
journey Home. Spiritual studies and meditative practices

have opened their minds to comprehend and accept a wider definition of truth. In their attempts to practice Universal Law in everything they think, say, and do, they are beginning to view themselves as consciously aware members of an evolved Earth community. They have made a choice to embrace Love's sweet vibrations to have and to hold forevermore.

Blessings upon you. You are a peacekeeper, an awakened eagle of the new dawn. It is our privilege to grant you an evaluation of our perception of humans that will, we hope, raise your estimates of your ability to reach into other levels of reality.

Our Task Is to Show You
the Way Home

The masses have erected a barrage of conditions intended to keep diverse human populations separated. In spite of their efforts to protect themselves by ignoring what is becoming exceedingly obvious, there is an underlying cooperative component to society that moves beyond the facade of exterior circumstances. Higher-dimensional beings witness all forms as elements of Light, however, and are privileged to view humans much more profoundly than humans do themselves.

Now that the end times are here, many are diving deeper into their interior to understand themselves as spiritual beings. By so doing they are discovering the key that opens the doorway to their more dominant Soul Memories.

Spiritual practitioners who uncover past and future lives eventually face the fact that they have spent many life-times occupying different sexual and racial bodies. When they time travel using hypnosis or creative visualization,

they often view themselves in a variety of forms and ages—
and not always as humans. They can't decide which body
derived the most pleasure from living. Ultimately, they
conclude that attitude, not physical structure or circumstance, determined how much enjoyment they experienced
in a life. Eventually, they determine that primary life
substance is Soul.

When the logical brain understands that Soul is unlimited by time or space constrictions, all fear begins to dissipate. Inevitably, there comes mental and emotional acceptance that Soul has countless opportunities to experience,
Self-discover, and evolve and that Earth is not the only
place to do so. Life now becomes much simpler. It is much
easier to quickly move through anxieties and self-doubts
when the healthy ego releases outmoded patterns that
keep the personality attached to unresolved karma.

Even a cursory knowledge of past and future lives is
sufficient for true seekers of spiritual knowledge. But this
information must be used to assist Soul's presently incarnated body to reach its ultimate goal—assuming, of course,
that this goal is to complete the journey and return to
Celestial Home.

Prebirth contracts for humans incarnating in end-time
energies have built-in clauses that relentlessly challenge
their Souls' third-dimensional aspects to awaken. Some
will. Some will not. Our task as trainers for galactic pre-
paredness is to present our human charges with a variety
of conditions that will provide opportunities for even the
most resistant to push themselves forward and upward.
Your guidance team has eagerly accepted the responsibil-

ity to outline the contours of the spiritual causeway and to provide you with an easy-to-read map Home. Beings of light who purposefully stay hidden in behind-the-scenes reality are your direct link to Prime Source Energy. Until ascension is achieved, it is essential for you to learn to intuitively interact with your angel-like guardians, for you are incapable of traversing the subtle stellar pathways without assistance from your etheric companions. Their only motivation is to assist and guide their precious charges until the bliss-filled regions of the evolved galactic community are reached.

No matter your personal circumstances—whether cosmopolitan or rural, wealthy or poor, unrewarded or well-recognized for deeds accomplished—we observe you as Oneness of Being. We perceive you as essences of concentrated light residing in dormant physical bodies.

To offset the effects of thousands, perhaps millions of years of spiritual stagnation, you must recharge your batteries by energizing your primary chakras. This is why almost all awakening humans concentrate on moving light through their spinal regions, why they are so intent on infusing their every cell and double-helix DNA with golden-white light. They quickly realize that a cleansing chakra shower is easy to accomplish. A simple meditative thought invoking the presence of Divine Light instantaneously infuses the entire subtle anatomy with purified energy.

Reprioritize your goals by appraising self as Soul harmonizing and co-creating with beings of light. Eventually, contact with these beings will become part of daily living. An acceptance of "phenomena" as natural occurrences helps the logic-loving brain to focus on something other than mundane matters.

Guardians have always been with you. There has never been one instant in any of your lives when they were not immediately available to you. Your task has been to remember to ask. That you have not always remembered has been all humanity's downfall. Nevertheless, though you toiled through life after life in darkness, Christ Essence as well as other beings of light periodically took physical form in the guise of supreme avatars to teach, encourage, heal, and comfort you. Our task is to remind you of your proximity to Divine Light.

Every Earth culture has been gifted with sacred teachings, written and spoken texts to encourage the inhabitants of a world shrouded in darkness to seek Light. The most basic of these teachings is so simple to comprehend that even the smallest child can easily understand it: to love one another unconditionally.

Granted, it is very difficult for the spiritually motivated to maintain themselves in a constant state of peaceful composure when dealing with the disharmonic conditions that permeate every aspect of human society. There is generalized ignorance of how Spirit actively interacts in human life. Most humans do not understand the dynamics of planetary evolution and the way gravity is connected to karmic law. Prayers requesting immediate relief from accelerating torments are increasing. The fear of loss and death is overwhelming the masses. The unspoken anxieties that saturate the thoughts of most humans are sending cascades of negative energies circling Earth's body.

Our task is to adjust the Earth energies that are not conducive to her celestial advancement. Thus, it is mandatory that large numbers of evolving and de-evolving elements within all species are removed from physical form to balance the unstable polarity effects of those energies. The distortions erupting on Earth, such as accelerating human conflicts and natural disasters, when analyzed from a big-picture perspective, are cosmically necessary. Because of spatial-time limitations and rigidly held concepts, most humans are completely unaware of the spiritual principles of a dynamic, living universe and, therefore, essentially do not see themselves as Souls with a cosmic goal.

Pleas for assistance, compassion, and mercy are more numerous now than at any other time in history. Humans are adrift on a sea of sorrow. They cannot see that Earth is moving toward a moment in cosmic time when her light-body will be restored. They cannot see that brilliant swaths of light are sweeping over her magnificent body and that all who live upon her have been covered with a protective coating of loving light. If this were understood, it would alleviate the suffering that permeates tender hearts.

To become consciously aware of the spiritual levels of the universe, you must direct a process of self-discovery inwardly. To uncover the essential beauty that resides underneath egocentric karmic debris requires a deep dive through the dark night of the Soul. Essentially, awakening has nothing to do with accumulating knowledge of extra-terrestrials, visiting power spots, or perfoming rituals— or even with psychic ability. These things are simply something to know and something to do. Awakening has

everything to do with accepting Self as Soul. When that
goal is reached, you will realize what an expansive being
you truly are. When you know without a doubt that your
natural form is effervescent star-quality light, you will be
able to move quickly through the vagaries of your physical
body's spatial-time setting. As an awakened being, you will
be able to dissipate any energies that hold your ego
thoughts in apathy and despair. You will rekindle your
flame of commitment to face life's challenges through acts
of compassionate, unconditional Love. You will recognize
the impermanent nature of temporal grief. You will align
yourself with synchronous timing whenever you find yourself in the right place at the right time; you will learn to
interpret the symbolic nature of repetitive events.

Our task is to assist you in shucking off restrictive
chains of gravitational karma so your unencumbered Soul
may soar freely through the heavens. We are helping you
construct a foundation from which you may experience
yourself as Soul in physical expression.

Intention

Blessed ones, your intention to radiate Love-Light's brilliant tones is transforming the planet upon which you live. Your will to be of service to Earth Mother as she completes her preparations to absorb high-plateau energy has been recorded in documents held within the auspices of the Office of the Christ. Your escalating levels of refined intention in thought, word, and deed are chronicled therein as evidence of your growing level of cognizant spiritual productivity.

We assure you, you may trust the processes of the events of your lives. It is paramount for you to know that you are loved and guided by entities from the highest levels of light whose intent is to fulfill the will of Omnipotent Source. Meditation is the most powerful tool at your disposal for overriding the doubt that continues to confront you. Retrain your minds to focus on images of expanding realities that are becoming available to you. Inwardly you can visit galactic ports of call that otherwise would be unreachable from your physical position in spatial time.

When engaging in creative visualization be very clear, remembering always that form is created from thought. You must be alert to what you desire to bring forth. Though the light of Earth's etheric body has grown much brighter since 1987, it will take a few more years for the last vestiges of darkness to completely dissipate from her exterior regions. As Earth enters her final phase of vibrational ascension, that which you thought-create establishes your level of participation.

It is a mistaken notion that godlike beings are constantly judging you and finding you wanting. *Observing* would be a more accurate word. That Which Loves All Things Unconditionally has long moved past cognitive thought. It is societal conditioning that attempts to make humans feel inadequate. Intention of thought, word, and deed are principal propellants that drive cause-and-effect events—not the dictates of temporal culture and law. Perceptions based on stringent religious doctrines stimulate little appreciation of the true nature of beings who dwell in Oneness with Universal Mind. Beings formed of components of living Light are best described as beings of focused, unconditional Love.

In these transformative times, the common people are having difficulty coping with the change. To compound their discomfort, a great deal of misrepresentation is being perpetrated by unenlightened humans who are becoming almost giddy with fear. Followers of dogmatic sects and organizations are doubly fearful. They point accusing fin-

gers at the more brilliantly lit and deem them false and
unworthy. It is almost impossible for those who are steeped
in fundamentalist religious and scientific systems to ascertain the true nature of those who have risen above their
lower psyches, for they are unable to comprehend the
motivations of truly spiritual beings. They suspect anything that even hints of psychic phenomena. They allow
themselves little or no freedom to explore the vast worlds
that lie within. Because of their fear of inward investigation, they have no basis to comprehend the multilayered
nature of greater cosmos.

Creation's inhabitants are made of the stuff of unlimited
potential. There is much more to the universe than even
the most enlightened human could possibly understand.
Most of it is grand and glorious. However, some of it is not.
Therefore, it is critical that newly awakening travelers move
through the unseen realms with discernment. Take careful
note of what you observe. Document your discoveries in a
journal. Begin to form a reliable accounting of your journeys. Although much can be gleaned from books, movies,
and other resources, your relationship with the universe
is predicated on the intent with which you approach its
vast contours.

If it is your intention to activate spiritual purpose in
the here and now as a member of the great intergalactic
task force that is ridding Earth of all negativity, confront
your doubts by creating a solid connection of Source
Power energies through your system of chakras and your
external auric body. Many awakening humans suffer from
bouts of depression. If you find yourself caught in cycles of

emotional ups and downs, make an all-out effort to prevent yourself from falling into the pit of critical self-assessment. Your emotional swings are due to much more than temporary personality adjustments and should not be viewed as necessarily negative. Not all that is affecting you is your own "stuff"! You are being bombarded with uncontrolled energy surges from the collective human mind coupled with highly refined cosmic waves that are interweaving human karma with Earth's fluctuating timeline sequences. A connecting of all your energy bodies—physical, astral, emotional, causal, and mental—is aligning you with the strained conditions that are pushing Earth into overlapping regions of third- and fourth-dimensional space.

As you know, Earth is in the process of fulfilling her Soul destiny—as are you. If your spiritual goal is a genuine, heart-triggered intention to ascend to light-body status, you are being thrust ahead at force speed.

Many of you are in relationships and jobs that leave you spiritually frustrated. We go so far as to state that in some cases the conditions are evil. Interaction with greedy, power-hungry, negative-exuding people is demeaning, counterproductive, and spiritually detrimental if it is not approached with high discernment. However, if you have made a solid commitment to Omnipresent Source, your circumstances are exactly in line with Spirit's primary intent: to spread light in all remaining pockets of darkness. Only the most courageous are uncompromisingly motivated to evolve. A great many have received assignments that require immense amounts of stamina and creative resourcefulness. These individuals, eagles of the new

134

CHRIST ESSENCE

dawn, exude energies of high moral integrity. They are brilliant dispensers of cosmic light. Their clearly spoken command to their higher Selves is the impetus that initiated the exact situation they find themselves in.

Many who present a serene and loving outward demeanor are inwardly tormented with uncontrolled rage, jealousy, greed, and hate. The faces of Soul-darkened individuals often imitate illusions of reflected light. Learn to unmask deception. If you find your daily tasks bring you in contact with people who are intent on deceit and betrayal, whose bearing is overtly sinister, silently send them a dose of illuminating cosmic light. Have the courage to speak directly to their higher Selves, asking that your gift of love be received on the inner planes as a surge of heart-awakening energy. Know they are twisted and torn from perceiving the world from a terror-based reality. They are starved for love and compassionate understanding. They suffer greatly. Discordant choices forced them into a posture of pseudo self-protection. Consequently, they have become proficient at pointing fingers at others and making false accusations.

The widespread spiritless behavior practiced by human leaders has brought the collective to a sorry state of affairs. Long ago the sensitives learned that to survive they would need to stifle their hearts' light and bury it deep within. Consequently, few are aware of the brilliant aura they project. The virtuous are the most self-critical. They routinely assail themselves with self-denouncements, judging themselves as failures and unworthy of critical notice. Nevertheless, from the ports of intergalactic starships we note that the most profound light issues from the surface of Earth. From our perspective, humans who reflect illuminated spiritual energy shine brighter than the flickers of electricity that pinpoint the whereabouts of Earth's major cities.

A Challenge to Love

The means to obtain personal knowledge of higher worlds is available through routine meditation. If your spiritual goal is to secure Christ-level consciousness and achieve light-body ascension in this lifetime, courageously break down the barricades you have erected to shield yourself from life's harsher realities. Uncover that tender heart you have coated to insulate yourself from all which is sad and dark. Awaken and observe your inner self as a bouquet of soft, multihued flowers. Awaken, and your beautifully budded Self will burst into bloom.

Trapped in the planes of negative-positive polarities, you have felt incapable of expressing unconditional Love as a constant force from which to draw life's energies. The intensity of your passionate feelings is at once uplifting and frustrating, a double-edged sword you have not always wielded wisely. Beloved unto the stars, your higher Self has the capacity to instantaneously transform pockets of Earthbound negativity into abiding Light. Having forgotten that side of your glorious being, as you awaken you struggle to regain that which you have lost. The Memories flicker at the periphery of your awareness, and for the most part, you have only a vague sense that your true essence is Spirit. Although you are beginning to realize that you have always been a member in good standing within the greater intergalactic community, it is not yet entirely defined within your emotional body. You struggle to remember that your Soul came to Earth with a specific mission.

In cooperative resonance with the Regency star council (your backup team) and other like humans, you have been issued authority by the Office of the Christ to dispense cosmic Light throughout Earth's grids. To effectively accomplish this task, you must maintain your patience and remain firm in your spiritual goals. Practice honing your intuitive skills and stay closely tuned to the spatial-time anomalies that are urging Earth through the multilayered, fourth-dimensional astral realms.

We are not unaware of your oft-spoken (telepathically) desire for a hasty conclusion to the processes that are pushing Earth higher. We caution you to move forward at intuitive speed; however, do not hesitate to pour copious amounts of Love-Light in laserlike fashion upon the Earth grids that seem to be unstable. In this way, you will truly assist Mother Earth in her efforts to reach the stars.

Do not be afraid and do not doubt yourself. We tell you, you do have the stamina to endure. You are an implanted starseed. You came to Earth from your Soul-star as an emissary of Light. You are an elite trooper in

Archangel Michael's great forces. Your task: to instill and anchor light wherever darkness is encountered.

End-time events are carefully orchestrated and are unfolding according to Divine Plan. In spite of dire predictions of a dying civilization, a new-dawn world will be erected on the labors of brave humans like you who have challenged yourself to awaken. Like a phoenix you are rising from the ashes of an outdated time to germinate the structures of an entirely new whole-Earth society. The inhabitants of this newly structured world, aware of their familial connection to the inhabitants of the galactic core, will live in peace, harmony, and unconditional Love.

You must be prepared to absorb and transmute the chaotic energies that can be observed in your escalating natural and human-triggered disasters. These things will continue until the sorrows of Earth's historic times are dissipated. There is a power within you that knows you have the ability to project Love and compassion to others. We encourage you to stand tall and firmly and unhesitatingly lend a hand to those less fortunate than you. Most humans are living lives of quiet and not-so-quiet desperation. They are frightened and confused. That which they have come to rely on is crumbling. Their world is changing so rapidly that it is becoming impossible for them to pinpoint areas of security. Stand firm and, as best you can when permission is granted, help others to understand that their true nature is Light and that there is no thing that need be feared. Your physical needs and urges will remain intact until Earth is solidly established in her ascended lightbody. Therefore, it is important that you stay alert and maintain a state of focused awareness. When the timing is right, your inner guides will inform you of the appropriate moment to unfurl your banner of Light and proclaim your starseeded connection. As the stellar winds whip your true nature into view for all to see, others may act as if you have suddenly become a pied piper, come to lead them to Light. When this occurs, you must move cautiously with an enlightened eye and sagacious wisdom.

It is imperative that the contents of this essay be carefully absorbed. Remain rock solid and grounded. It will benefit no one, least of all yourself, if you take this information and run with it from a position of unbalanced, glorified ego. The intention of these data is to encourage a practical approach to life as an awakening Soul. Cosmic power is dispensed by beings who are grounded in humility, compassion, and unconditional Love.

It is time for you to think of yourself as celestial-magnitude Soul. When you have Self-activated, your multi-hued energy bodies will begin absorbing and transmitting stellar Light as if you were a miniature sun. You must wield your newly discovered powers with impeccability! To help you accomplish this astonishing feat, keep your inner ear tuned to the cosmic hotline for those beings who represent you in the star councils and in the Office of the Christ.

After you learn to function by the same standards as do galactic citizens, you will be introduced to a multitude of beings from many star systems. You will communicate telepathically with upper-realm entities. Your first step, however, is to practice artful, spiritual interaction with other forms of life on Earth. Strive to mimic your extrasolar companions' way of life.

Carry your personal torch high. Although it may appear that light grows dim in these troubled times, know that in reality Earth's etheric bodies have begun to illuminate as brightly as high-magnitude stars.

Follow me. I have come to banish Earth's enemies into the recesses of a dismal past. I engulf my beloved, loyal followers with Light and Love. You may know me as Jesus, Buddha, or Krishna. You may know me by a thousand other names. That by which my I Amness is identified is not an issue. What is vital is your devoted adherence to your spiritual path and to the manifestation of unconditional, compassionate Love on Earth. These simple instructions have been made available to all humans. They exist in the sacred literature and teachings of all cultures. To you who are opening your hearts and minds to practice living as spiritual beings, that which is promised will eventually be realized. Remember, these are the holiest times humanity has ever known.

Prepare to Receive Light

Beloved readers, whether you approve or not, Earth has always been carefully scrutinized by otherworldly beings. Human activities in particular are monitored at all times. Many are aware of our presence. Those who inform others that we are here suffer ridicule. The message that would transform human society is scorned as nonsense. Most will not acknowledge the possibility of extra-terrestrials, let alone angelic beings and masters of light. They believe that to speak of such a thing would precipitate a mighty invasion. It is as if an enormous "Not Welcome!" radio signal were being broadcast into space. Humans fear losing what little they have. Those who are certain of their assumed freedom to do exactly as they please are afraid they would be required to forfeit their right to dominate and control the less fortunate—which, of course, they would.

What if we were to suggest that all it would take to free you from your bonds of fear is a large dose of applied courage, that you have the power within you to rebel against the pseudopower of unsaintly authority? If you would do so, individuals and events that only appear to hold you captive would naturally go away and never bother you again.

Rise up from those uncomfortable wooden pews and solemn, joyless sermons! Run outside and look to the sky. By day and by night, from horizon to horizon, oval-shaped clouds announce the presence of interdimensional starships. The skies are full of their brilliant hues, yet the masses go blithely about their busy-ness in complete ignorance of their familial relationship to the rest of the cosmos. Nevertheless, the arms and hearts of the more attuned reach skyward with pleas of "Home! Take me Home!"

How long will it be before humans respond to the gravity of their situation? How long before the masses passionately dedicate themselves to assist the angels in implementing Greater Plan on Earth? The end times are upon you! Whether acknowledged or not, everyone feels the energies from the shutting down of third-dimensional history. That which transpires is in the latter stages of completion. The final wrap-up is just around the corner. Humans are being challenged to mature spiritually. How do you suppose the uninitiated will survive massive change if they fail to honor the proximity of Archangel Michael's mighty troops and the beings of light who stand ready to extend Divine Grace to those who embrace Earth's evolution to Light?

Energies sweeping over Earth originate in the hallways of the Central Sun. The waves are so powerful that the

143

PREPARE TO RECEIVE LIGHT

individual has little chance of absorbing and using those energies without angelic assistance. To activate Divine Intelligence in your life in a purposeful, integrated, co-creative way will require your act of intentional will. You must, with every fiber of your being, opt for and prepare to receive high-resolution light. And if you have not already done so, you must establish a telepathic link with your inner guides.

The brotherhoods of light and the Office of the Christ are quite aware that these are agonizing times for you. Your civilization is being plagued by catastrophes and bombarded by vast amounts of energies from outer space. You are feeling vulnerable and unfairly jeopardized. The illusive contours of your reality are slipping away. In a desperate attempt to regain some footage, many have challenged themselves to awaken. Seeking to obtain some sort of blissful composure, they are finding it difficult to do so when confronted by the fear-filled faces of the slumbering masses.

We cannot stress too emphatically that you must learn to follow your intuitive heart's wise council. Meditation focused on integrating Love-Light into every pore, every cell, every genetic helix of your body is necessary if you intend to ascend. Meditation is your primary key for transformation to Light. Meditation techniques will help you develop a clearer understanding of what is taking place as you collect spiritual knowledge. Open yourself to greater possibilities. The ability to keep an open mind will serve you well. Strive to be a living example of forthright goodness. Practice kindness and mercy toward self and others. Put into action that which you profess to believe.

Although your battered ego may be feeling as helpless as a newborn babe and emotionally stretched to its limits, never stop yearning to intimately know God and the great beings who inhabit the light-realms. Enthusiastically embrace nuggets of cosmic wisdom when you stumble upon them. Tend to your daily busy-ness from a place of compassionate, sensitive awareness. Do not befuddle yourself with inappropriate egocentric indulgences. Attempt to understand at every level of your being that you are Soul, a necessary cog in the massive rotating wheel called uni-verse. Come forth and majestically proclaim that you are a being who honors its connection in Oneness with all things.

Initiation

I come to invite the spiritually discerning to petition Spirit for the opportunity to participate in celestial initiations that will propel them, fully aware, into realms of remarkable Light. To facilitate clear telepathic reception from beings who already dwell in universal harmonic bliss, you must cleanse the pores and cells of your physical bodies of all negative residue.

In your current life you are undoubtedly experiencing difficulties on many levels: financial, relational, physical, emotional, mental, and spiritual. In a very real sense, your entire lifetime, from birth to death, can be regarded as a test of your intention to maintain yourself in a state of integrity and to become karma-free. To this end, your higher-Self guidance team is kept busy providing windows of synchronistic opportunity so that centuries of repetitive Earth cycling may be completed in this lifetime. Having been trapped by the gravitational pull of a lower-toned planet, you are now making a valiant effort to relieve yourself of cause-and-effect burdens created from events, places, and people your conscious mind has long forgotten.

You should regard every aspect of Self as a classic work of art created by the all-encompassing mind of Supreme Celestial Artist. It is not enough to merely observe a magnificent sculpture from a distance, but some cosmetic dusting and restoration must be undertaken from time to time. So it is with the human creation. To assist your unseen companions in their task of helping you sweep your energy bodies clean, become aware of every molecule of sensation that moves through you. Remember, to know Self utterly is to know God profoundly!

The inner craving to Self-discover spiritually is not egocentric. To honestly observe every aspect of yourself, to acknowledge and understand every nuance of energy that moves through you, is an act of humility. It requires absolute trust and faith that the universe is a loving place and a knowing that its urge is to sustain and abundantly nurture your every need.

By nature, spiritually motivated individuals are geared to attain conscious communion with all universal beings, to become One with the entire cosmos. The effort it takes Earthbound beings to rise above the heavy weights that keep them tied to a third-dimensional planet requires a journey through the ego's darkest aspects. To move through and then rise above the night of the lower astral that surrounds Earth's spatial time demands tremendous effort and courage. One must be forthright, persistent, and determined to remove every oppressive constraint that holds one back.

Much of that oppressive substance is an accumulation of karmic-related material—thought forms and other sludgelike negativity that have piled up over many lifetimes. In addition, demonic entities inhabit the lower-astral regions. To move through the roadblocks they will set up in an attempt to deter you from your path, you must command Light and learn to use it with laserlike precision.

Clumps of unvalidated past and present traumas are stored in your body's cellular makeup and are buried deep in every double-helix DNA. They are almost screaming at you for recognition! Your unresolved issues are your personal markers for spiritual evolution. They are guideposts that point your way to the stars and Celestial Home. In earlier times the curtain surrounding Earth was much heavier. To achieve liftoff, spiritual initiates were required to sit in constant meditation and work directly under the watchful eye of a God-realized master. In some cultures, priestly initiates would be required to spend time in tomblike enclosures—pyramids, mountain ranges, and other shadow-filled places. There they would be forced to look directly at the face of their own dark nature. Because the webs of energy that surround Earth have been loosened, such tormenting measures are no longer necessary. Now spiritual initiations are carried out within the framework of your daily lives. Master-level spiritual teachers are available only to the few; determined individuals working in telepathic harmony with their unseen guides can achieve transformative results. A constant that remains in effect, however, is the need to petition Earth's Spiritual Hierarchy, the Regency star council, and the Office of the Christ for assistance in graduate procedures necessary for ascension.

You must cultivate an attitude of spiritual integrity. There is little time remaining to do so. Get those self-growth legs moving! There is no time for delay! Few years remain in the course of human history before interdimensional beings land their glimmering starships and invite you to join their celestial celebration.

Hear me, O reborn ones! You are hybrid elohim reestablished. Lords of Light, Love, and Harmonic Splendor beseech you to assist them in reinvigorating Prime Light into every facet of Earth's contours and into every illusive nook and cranny of your being.

Hear me, O latter-day humans, for I have come to you through these pages to invite you to the stars. In the Light and Magnificence of the Omnipotent One, I Am! I Am!

On the Jupiter-Comet Event

This transmission is directed from the Pleiadian star system, transmitted expansive mode via our Sovereign Leader, Sananda, the Christ Essence. Direct linkage is to the universal state of Nirvana, home of realized Omnipotent Source in the form of the One who provides access for spiritual-striving humans to achieve cosmic citizenship via light-body ascension.

The star councils invoke praise on you who are aware of the rapidity with which Earth's planetary brother, Jupiter, is evolving. Earth's restructuring is reflected in mighty Jupiter's ability to absorb solar-grade light in a manner not heretofore experienced.

Many a picturesque scene was captured by Earth-based cameras as comet fragments plummeted onto Jupiter. Alert to the possibilities of dramatic structural changes on Jupiter, many scientists secretly wondered what it all meant. They were both fascinated and appalled as their imaginations were uncomfortably tweaked by mystical Soul images that threatened to break into their conscious minds. Fortunately, this major astronomical phenomenon took place far enough beyond Earth's spatial-time placement that they were in no danger of premature exposure to such sweeping galactic changes—that is, until they are better prepared for such an unseemly encounter with space debris.

Though this transmission will be published some years after the 1994 Jupiter-comet episode, it may still serve as adequate warning that what transpired on Jupiter could very well preview a like incident on Earth. Be also informed that as you actualize as Soul having a human experience, doomsday prophecies and traumatic events will cease to cause you undue alarm. Keep in mind that fear is a tool disruptive entities use to divert the attention of the masses from achieving recognition of their true Selves. The spiritually sagacious understand there are profound underlying reasons for the disruptive-appearing, chaotic occurrences taking place within this planetary system. The situation on Jupiter was indeed a prelude to escalating natural and human-caused disasters, for while Jupiter was pounded with comet fragments, a great deal of energy was released into the solar winds.

Earth is not the only planet where evolutionary changes are taking place—although, to you, she is the most obvious. Mars is experiencing a realignment of its etheric bodies to a level parallel to several other planets, in particular Earth, Jupiter, and Saturn. For the most part, the scientific community is not yet aware of the etheric planes and the vibrational adjustments occurring within and outside this solar system. However, people who approach universal mechanics at the quantum level are beginning to observe a fragmentation of consciousness that is merging into holistic Oneness.

Divine Opportunity has issued an invitation to the beings who inhabit this solar system to heal old wounds plaguing them since ancient times when a great war was fought. That war culminated in the abrupt rendering of the planet Maldek into a billion pieces and the formation of an asteroid belt between Mars and Jupiter. Those planets, being physically closest to this doleful event, have suffered greatly from the aftereffects of Maldek's destruction. Although Mars remains barren, Jupiter's enormous energies flare with the intensity of a birthing sun. So be it.

Fragments of cosmic matter diving into and disrupting Jupiter's repose are reminiscent of ancient warring energies that have periodically devastated not only entire human civilizations but inhabitants of other planets within this star system as well. Under more harmonious conditions, third-dimensional populations would be thriving in Eden-type environments on Venus, Mars, Maldek, and Earth. Sol's planetary family, living beings that they are, almost gave up in despair a long time ago. They deeply grieve from being treated with such little regard by entities whose very existence they made possible and supported.

Withdrawing into herself, Venus made a valiant effort to rid herself of all third-dimensional life by causing her surface to be covered with poisonous gases. Simultaneously, she pulled up her evolving occupants into her fifth- and sixth-dimensional regions, where they still exist. For a very long time, Venus's upper domains have been home to the most highly evolved beings in this planetary system.

If you explore history from a metaphysical point of view, you will uncover indications that Earth has periodically undergone abrupt and rather radical physical alterations. At one time she was bombarded with large amounts of cosmic debris and suffered a near collision with a planetoid body. There is much information available regarding these traumatic events, and you are advised to search them out.

The de-evolving portion of the human family is on the brink of precipitating an unparalleled tragedy that could instantaneously devoid Earth of all life. However, those who are awakening to the true situation are beginning to understand that they are witnessing a transformation of time itself—of energies of unlimited alternative futures. Those who set their sights high enough can see the outlines of an entirely transformed human species.

We are aware that overly preachy entities with little understanding of what it means to be current-day humans are attempting to force-feed extraordinary amounts of unsolicited information onto the planet. Most humans feel they have quite enough to deal with without pushy multidimensional beings meddling in their affairs. If humans had tended to their planetary responsibilities in a wise and respectful manner, however, outside interference would not be necessary. Nevertheless, the damage humans do to themselves and to Earth's environment is the business of all galactic citizens. The acts of aggression you commit against yourselves are acts of aggression on all beings, for there is no thing that is separate from another. Thus, the intergalactic crews are busy sounding warnings throughout all human communities. Our imprint grids are reflected in the growing acceptance of extraterrestrial visitation and the popularity of UFO sites on the Internet. The World Wide Web is connecting your information banks into unprecedented Oneness. Subliminal transformative stellar energy bites are exuding from all Earth-based computers used by corporation managers, stockbrokers, secretaries, accountants, and architects as well as from millions of personal computers in households all over the globe. Extraterrestrial communicator chips are embedded in computer cores of all governments and news organizations to monitor and prepare for eventual upgrade to stellar-level capacity. The latter will occur instantaneously throughout Earth's electrical grids when purposeful multidimensional contact is made.

Humans who interact with negative extraterrestrials are very proficient at scrambling electronic data in their attempts to control the minds of the masses. They are not limited to computers but use radios, television, and so forth. Electronic devices are as easily accessible to beings of light, but as of yet, data based on Love-Light teachings are not as widely accepted.

Are you beginning to feel mind-invaded? Despite your discomfort, this is not a new thing. Before the advent of computers, your radio wave–accessing technology tapped into Earth's light-enhanced power grid.

If you even barely grasp the ramifications of this essay, its implications will probably set your teeth on edge. It is important that humans begin to understand: Events that took place on Jupiter in 1994 began an influx of powerful cosmic energies into this solar region. We challenge you who are endeavoring to awaken to stay tuned to Light and to envision Earth transformed into the contours of a light-emitting planet.

As is Earth, Jupiter is being prepared for ascension. The comet fragments provided a boost of enormous power for this purpose. Jupiter is a sun in the throes of birth—not an act of cosmic aggression but of ascension into higher orders of God-Recognition. Achievement of high-resolution Light by a planetary body is an awesome thing. Congratulations are in order for what Jupiter has already accomplished. Jupiter is paving the way for all planet bodies within this star system.

Adonai.

Complete Your Cosmic Homework

It is not the purpose of the Office of the Christ to thrust new assignments on you. That which was anciently set upon the planet remains intact. That which was outlined as basic to all spiritual dialogue is retained in the growing interest in metaphysical concepts and in renewed interest in philosophical discussion, alternative methods of healing, and fundamental ethical principles of moral behavior.

Popular trends in psychology are simplifying the rigid edges of this scientific discipline, and a multitude of books on the subject are within easy reach of most humans. There is something for everyone, at every level of reading and personal comprehension. This is a very good thing, for what was once coveted and kept in the private domain of highly educated experts is becoming accessible to everyone.

Why is this so? Because the first phase of your homework is an in-depth, honest analysis of your ego patterns. The rapidity of your spiritual growth is directly related to your ability to develop a working knowledge of who you really are.

The next phase of your assignment is to courageously challenge yourself to examine your dark side—to critically identify and acknowledge your personality's heavy tones, that which holds you tied in karmic-laden habitual patterns of anger, frustration, jealousy, greed, pride, laziness, and so forth. Using techniques you have learned for focusing spiritual light, transform areas of stubborn resistance within you until your energy field shines as brightly as a birthing star.

Throughout these writings we have showered you with the pleasant news that you are a beloved member of a family of multisystem stellar beings whose very presence on Earth is a given. This is something new in the human experience. From time to time we interrupt your reveries to express our gratitude that you are developing an increasing interest in personal growth. We are best able to help those who help themselves—a maxim particularly true in today's world.

We cannot help humans provide a peaceful planet upon which to live when they insist on remaining in the throes of interpersonal, regional, national, and international conflicts. Because attention to Self-transformation is not an across-the-board effort, for several years the human species will continue to suffer the pangs of violent, destructive events—a necessity for Earth to complete her healing journey.

To best assist you during these difficult times, beings of impeccable light reach down from their berths upon rainbow-hued starships to provide you with their special brand of cosmic nurturing. No one who petitions for ascension is without an entourage of light-evolved guidance counselors.

Further into these dialogues we will introduce you to the electromagnetic circuitry of unfolding alternative timelines and to concepts regarding your innate ability to tap into multiple realities (see in Part IV, "Grid Conduits, Emerging Timelines, and Alternative Futures"). The quantity of grid conduits housing future Earth events is essentially unlimited. Your ability to time-track into more refined webs of energy is predicated on your willingness to move beyond third-dimensional and even fourth-dimensional time restrictions.

Graduate-level homework is based on understanding the harmonics of now time as essential to stellar-grid navigation and higher-octave operational parameters. Illogical as it may seem, beings of light experience time differently than you do. Past-present-future holds no meaning above fourth-octave level except in observational mode when negotiating with beings such as yourselves.

To enhance your homework assignments, we are presenting you with many maplike outlines that will help you pinpoint various time scenarios suitable to your personal path. Occasionally, various webs will draw you down several pathways at any given time. Your etheric counselors stay alert to your changing needs and continued quasi-sleep condition to help you ascertain the most suitable channel to follow. They recommend this or that as they tirelessly present you with mainline and optional route conditions.

For the most part, their advice will not be given when not solicited. When you reach critical junctures, however, you may feel a not-so-subtle boost in a certain direction. Patricia has coined these jabs the "celestial crowbar effect."

We cannot emphasize too strongly that, to us, you are beloved cosmically innocent children. As parents true, we provide you with everything you need to open gateways of time based on synchronous moment-to-moment events. It is our task to provide you with materials and tools that will assure your graduation from Earth school. It is your task to recognize them and put them to good use, remembering that higher-level skills function best when approached from the intuitive mind.

As a graduate student you are provided with many detailed, multilevel training manuals. Nothing will be accomplished, however, if you do not put into practice what you have learned.

Perhaps the goal to become as light as an angel seems too much for a mere mortal such as yourself to aspire to. In your more self assured moments you grant that ascension is the ultimate reward. To accomplish this worthwhile feat, you must learn to integrate Light's essence into your emotional, physical, mental, and astral bodies. This is best accomplished through methods of creative visualization. A simple enough task, but time must be allotted to pursue your advancing studies in a meditative mode. That which is experienced in stillness must appear as real as an excursion into the world at large.

You must do a great deal more than dream about ascension and discuss its possibilities with your friends.

You must make it happen by completing your homework assignments. This is not a project that can be foisted onto others. Keep in mind that your Soul had a specific reason for choosing to tackle Earth-school curriculum. What you must come to terms with is if it is worth the demanding effort it will take to pursue Soul's ultimate goal.

Final assignment: Masterful perception of and ability to practice Law of One.

Graduate goal: Ascension.

Starseeds Becoming Light

Starseeds becoming light, stretch upward to meet the stars. Resembling high-wattage lamps, your activating etheric bodies radiate stellar-level light from Earth's surface. You are beacons of the dawning new age. You stand ready to access the portals of fifth- and sixth-dimensional starships.

In the not-too-distant future you will find yourselves teleporting from planet to planet and from star to star. As aware galactic citizens, your inalienable right will be to do so. Your long term of servitude to Earth draws to a close. You are almost ready to return to your originating Home—to Arcturus, the Pleiades, Sirius, and other wondrous ports of call.

Retain patience with those who are less knowledge-able than yourself. We do not praise you to promote a swelling of ego superiority. Remain balanced and com-passionately humble before the suffering of those who lack awareness of the processes Earth and all her inhabi-tants are going through.

Now that you have begun to get a glimpse of how very vast God's kingdom is and the countless mansionlike planets and intelligent beings that are "out there," you have begun to dream dreams and see visions that heretofore escaped you. You have reached the conclusion that access to the heavenly gates is an inward journey which takes place during life. You understand that many of your closest Soul family members are beings of light. Is this not exceedingly good?

Many will not ascend when the starships come to Earth. Vast numbers of humans have very young Souls and are not yet evolved to the point when they can grasp concepts of Universal Law or think of Planet Mother as a living being. They are even further from comprehending themselves as Souls having awesome human experiences. Many young Souls by necessity will travel in the ways of darkness, for the pathway through the lower realms is one that must be taken by developing Souls. In the galactic sector of which Earth is a part, lower-level darkness is particularly dense and potentially terrifying, unlike less warlike regions of the universe.

Older Souls, members of the starseeded teams that first implanted Earth, are awaiting transdimensional evacuation and elevation to previous-state harmonics. It is time for the Return. They have assimilated all that Earth school has to offer. It's time to be off to experience other challenging episodes in Soul's long journey Home. This is exciting news, for they are anxious to be on their way and are preparing with gusto for this exciting turn of events.

This is your call. It is about living a life devoted to the precepts of Universal Law. The world is poised for chaos. All is astir. We ask that you refrain from scampering away to a place your intellect deems safe. Face the turmoil head on! Go forth and brush the wounds of your fellow humans with ointments of gentle understanding and softly spoken words. Caress their pain with tender hugs. Take their hands in yours and tell them they are beautiful. Approach everyone with an open mind. Learn to withhold undue criticism and judgment. Be courageous and maintain a cheerful attitude.

Humans who are drawn to these celestial preparation manuals are the ground crew of a great intergalactic fleet. As you took up residency on Earth you were put in a state of Soul amnesia. Now the Memory blocks that were inserted when your spirit body first arrived on Earth are being lifted. Bit by bit you are beginning to remember why you are there and what you came to do. Most of you are attempting to make contact with your etheric backup teams, who remain eagerly at your disposal.

Sooner or later you must accept that many Earth humans will not achieve full light-status this time out. Breed compassion and not contempt for those less fortunate than yourself. In no way is a young Soul of lesser quality than an old Soul. Young Souls are not fallen angels. They are simply unprepared, untrained Souls who have many lessons to learn. Would you consider a wee kindergarten child as being of less worth than a soon-to-graduate college student with a doctoral thesis? Of course not. The analogy is the same.

The call is out for all frontline celestial troops to dig in and persevere. You are needed where you are doing what it is you do. Your instructions are clearly engraved on the contours of your awakening hearts. You will be of great assistance to your etheric backup teams if you carry out your assignments in a timely manner and to the best of your ability. We will supply you with all you need to fulfill your missions. Your primary tool is unconditional, compassionate Love dispensed with laser-Light precision.

The Shepherd and the Angel

Many millennia ago, in the early days of developing humans, starships descended through winds that stirred the sands of ancient lands and hovered above Earth to observe all that transpired below. Some years after their arrival, a lonely tender of camels and sheep emerged from the shimmering desert wastes that surrounded the little town of Jericho. Except for his flocks and a steadfast belief in the presence of the One God, the shepherd wandered alone, forever seeking he knew not what.

On a starry night some weeks before, the arid desert had abruptly cooled when a storm swept down from the mysterious northern realms. Shrieking winds drove dark clouds through the normally placid desert sky. Bolts of lightning streaked Earthward. Sheets of rain replaced the oppressive heat.

Suddenly, from out of the clouds the fiery figure of an angel appeared. His rapid descent brought him to rest before the terrified shepherd. Settling comfortably onto a sandy protuberance, the angel, in soft, pleasing musical tones, spoke to the shepherd:

"Hark, my good man. It is recorded in the books of Heaven that it is your way to gather flocks of sheep and camels and to ceaselessly herd them across the desert floor. For many years you have tenderly and dutifully led your dependent charges from one cool oasis to another, to places where water flows in abundance and green, nurturing grasses grow. Never complaining, you wander from town to town until you find them shelter and sustenance and loving people who are willing to care for the older, weaker members of your beloved herd.

"It has not escaped the attention of the star-masters that it is also your way to transport goods across vast desert distances. This simple service provides access for many diverse people to communicate with one another and to enjoy one another's fruitful bounty.

"Come now, dear man. We are quite aware of your gentle manner and great generosity of spirit. Because of your unwavering devotion to service, I have come before you in the storm's heightened energies to ask you a question. I would have you explain to this angelic entity the reasons for your unwavering passion to know and serve the One God. Although as an angel I am quite familiar with every aspect of Divine Presence's Omnipotent Being, I am curious why a mortal such as yourself is so insistent on evoking images of the One God in your mind's eye."

The humble shepherd replied, "O Great Winged One, I was born from the seed of my father joined in blessed union with that of my mother. Growing into manhood,

I left the peaceful sanctuary of my boyhood home only to discover myself trapped in a land of suffering and sorrow, of hardship and laughter, of love and loneliness, of life perpetually begetting death and death perpetually begetting life. Caught in a disturbing awareness that life is simultaneously joyful and miserable, I earnestly sought to intimately know Most High God. Saturated with life's bittersweetness, I hungered for His unwavering presence. I yearned to believe beyond any doubt that no matter what form my life took I would always be guided and protected."

Delighted with the shepherd's innocent response, the angel continued. "Blessed are thee among all men, for without the aid and comfort of a teacher you have struggled alone to understand the depths of the covenant that exists between God and humans. Further, I state unequivocally, you have always been accompanied by angelic presences. As I am beside you now, I am beside you at all times. Though the majority of men and women who inhabit Earth will pay scant attention to us for thousands of years to come, the sun will eventually dawn on a brighter day, a time when humans and angels will communicate one-on-one.

"Within the core of your eternal Memories you will recall the night you spoke with God's angelic messenger. After the energy of this life is extinguished you will return again and again to human form. Your most important life will be lived far in the future when you will take on the form of a small, froglike man. You will be wrinkled, gnarled, and exceedingly wise. When you are very old you will be a great leader. That life will take place in a land that lies a long way from these sandy desert mounds."

The humble tender of flocks sat in awe as the angel told him of magnificent worlds that exist in the far-flung, starry heavens and of the great brotherhoods of light who guide initiates into the sacred realms of Divine Knowledge.

The next day, after the angel had ascended, the shepherd gathered his charges and proceeded toward Jericho, where he planned to gather supplies for the long months ahead. As the noonday sun burned its way across the desert he came to an oasis of cool, clear water. Leafy palms stretched high overhead. Heavy with the sweet odor of ripened dates and succulent figs, their branches cast refreshing shadows over all who sought refuge from the heat. The shepherd's mind rested easily now, for he knew in his heart of hearts that angels are always beside him. He could feel the soft caress of their wings.

Late that night as the shepherd peacefully slept, the angel appeared in a dream and the humble shepherd continued his growth toward perfected wisdom.

In the dream the angel spoke to him again. "Blessed are thee among all men. Before your dreaming eye lies a golden land of opportunity, a place where fruits and nuts are plentiful, where delicious honey trees abound and streams of sparkling water run free over verdant fields. In spite of all they have been given, the people of this land suffer greatly from pangs of terrible loneliness. They are a miserably frightened people, for they have separated themselves from one another because they lack belief in the One God and the existence of angelic beings and ships from the stars."

Observing the landscape with his dream eye, the shepherd saw a bewildering display of immense cities composed of unimaginably tall buildings. He saw a multitude of devices that moved swiftly over roads without the benefit of oxen, and he saw boxlike cylinders that flew through the sky with outstretched arms. He was astonished by those strange sights, and a great fear began to overtake his dreaming mind.

Offering the shepherd little comfort, the angel continued. "It comes time for my departure. Before I fade from sight, I leave you with this parting remark: Some thousands of years hence, you will be born into the land you have been privileged to envision. In that lifetime you will be a primary caretaker of sacred knowledge. You will enter that body with Soul Memory intact. As you grow to manhood you will find yourself in a world plagued with a profusion of confusing teachings. Because of your clear grasp of celestial wisdom, coupled with profound teaching and healing abilities, you will be much maligned by bishops, priests, and other proponents of shallow-based dogma. Because it will be a time of great darkness, an ever-present fear will cover the land like an enormous blanket. The humble in nature, those whose lives are devoted to the higher good of all sentient beings in service to Supreme Intelligence, will know you well, but they must speak your ancient name in hushed, reverent tones. Because it will not be your way to cower before another, you will approach every interaction and every circumstance with an attitude of joyful and compassionate but firm loving-kindness.

"Your principal task in that lifetime, which will be during humanity's greatest peril, will be to actively assist the brotherhoods of light in locating, protecting, and guiding those who are called to awaken. Your pledge to be a humble shepherd and conveyer of primary goods across lonely desert wastes in this lifetime is to initiate you into an understanding of these important functions. Most ancient of Souls, you agreed to set a preparatory thought form in place when you pledged to gather and attend camels and sheep in the birthing years of what will come to be known as the Holy Land. You already have had several incarnations in these lands, and you will come again and again. Humanity's sacred records will write your name 'Abraham, Father of Nations.' One of your primary tasks will be to act as father, teacher, and guide to one known to history as Jesus, the Christ. You bear the courage and humility to step aside as He is christened the Enlightened One, the Redeemer, the Shepherd, the Lamb. Never, since the beginning of humanity's sojourn upon this planet, will you falter in your eagerness to serve Omnipotent Radiance.

"In the epoch of the starseeding, already a Soul of great antiquity, you journeyed to Earth—the planet that star-masters refer to as the Plains of Saro—from the stars. Now, as you wait for the distant age to come, you will drift and dream through many lifetimes. In some you will busy yourself as a collector and seller of merchandise, in others as a creator of fashionable works of art. You will scrape hides with bones, cook, clean, and bear children. You will work with plow, saw, awl, chisel, rudder, and mast. You will become a great chief, a leader of men, a signer of important documents. Each incarnation will constitute a critically important step in preparation for your Soul's greatest work—as one who gathers and leads awakening lambs to Light.

"Simplest of men, you are wise beyond time itself. You carry within your loins the seedstock of nations. Ironically, from out of your primary goodness a world of people will germinate and grow as a cell divided. A frenzy of misunderstanding will separate one nation from another. As a result, the people will experience the dualities of joy and grief, peace and war, serenity and almost overwhelming confusion. They will writhe in fear when confronted with disease, jealousy, anger, doubt, and hate. Yet, from out of this seething chaos a gentle people will arise: the Awakening, the elohim reborn.

"Earth's Spiritual Hierarchy has deemed you most capable of gathering the starseeds and of dispensing information they will need to sustain them until the moment the Dove descends, an event scheduled for the twenty-first century following the death of Christos, known as Jesus to His beloved followers. As the twenty-first century comes into being, an Era of Radiant Light will be born.

"Now you must settle yourself in anticipation of the long centuries that lie ahead. Blessed One, I Am Uriel, of the High Arch of Heaven. Adonai. Adonai."

Blessed are they who approach these writings with an open mind. May the wonders contained in these Arcturian-energized essays provide you with the wherewithal to stay courageously watchful and attentive. Know this: The covenant between Archangel Uriel and the Shepherd Father originated in Beyond the Beyond, in the domains of

Celestial Sun. May hope and faith sustain and nurture you during these difficult times. When things seem the darkest, do not forget that night is the precursor to day and that the brightest light Earth has ever seen is about to burst forth from the sky.

This tale has been transmitted to assist your spiritual efforts. Keep in mind that these pages are embedded with high-magnitude energies designed to expand your knowledge and mastery of the enlightened, Christlike state that is natural to Soul's essence.

173

Titanic Beings of the Galactic Core via Sanat Kumara serve the Regency star council for Earth's propulsion to light.

Titanic Beings are securers of galactic stargrids to Source attachments radiating from Central Sun. They exist both outside and inside vibrational octave realms. They emanate Divine Light from the Home of Universal Central Sun.

Poem

We know of your suffering and
    we have made it ours.
We have entered Earth's nether regions
    for the purpose of transposing dark to light.
        To establish principles of Universal Law.

Thus we have spoken, and it will come to pass.

We are the Omnipotent Ones.
    We are ones who bind cosmic Soul
        into One vast being.

Name the Star!
    We are that!
Name the Being!
    We are that!

We come from Beyond the Beyond,
    We are the Titanic Beings of the Galactic Core.

Introduction

Preparation is underway for an influx of energy in the form of exuberant light to override third-dimensional planetary timeline grids. On a personal note, incoming cosmic energies may be used to amass the fortunate things of life—that which is exceedingly good. It is a time for gathering the fruits of universal abundance.

By disbursing light from our position in the galactic core, sustainable vibrations are propelled in a crisscross fashion through space. Tides of wavelike energies weave and blend as they thrust forward, engulfing the spatial fields of the evolving universe in an uplifting movement. At this time, discourse will be kept to a minimum. The Arcturians have provided you with material basic to this theme.

To assist your transitional efforts, several galactic beings are joining your primary guidance teams. Our simple task is to expand the heart chakras of starseeded individuals in a stellar-grid configuration to mirror the essences of their home suns. As Earth is buffeted by radiants emitted from the core of the Central Sun, an astounding side event is taking place. Your lotus blossom heart chakras are glowing with effervescent light. Hereafter, the expressions "gratitude" and "good will to all" will be reflected in an interconnected solar-region embrace that is enhancing the movement of growing Oneness, essential to developing humans who will populate the new-dawn world.

177

INTRODUCTION

Creator's Building Blocks

Waves of Creator's formative energies are flooding Earth and the solar system in which she resides with refined light emanations from the galactic core. Vibratory sound-light surges are sweeping through the entire Milky Way galaxy. If viewed from the human standpoint, the reasons for this display of cosmic energy would appear more symbolic than actual. That is, the sleeping masses are unaware of the vibrant nature of the stars or that their primary purpose is to intone the song of the stellar chorus. The telling of such an outlandish tale would, therefore, appear to be irrelevant, part of the modern-day mythology proclaiming that Earth is being visited by extraterrestrials.

Nevertheless, Earth's more aware inhabitants intuit through every fiber of their being that vibrantly textured cosmic sound-light waves are modifying the contours of third-dimensional space and are dramatically altering linear timeline sequences. Those who investigate metaphysical—so-called channeled—solar teachings know that all particulate matter and nonmatter elements contained within this multilevel galaxy are either being uplifted a note or two or are moving downward a note or two on the universal harmonic scale. Concomitantly, because all things affect all other things, the entire universe in which the Milky Way galaxy resides is being inundated with swirls of vibrationally fluent sound-light energies from which Creation's building blocks are formed. These ribbonlike streams incorporate that which lies below as well as that which lies above. As each multileveled dimension moves either upward or downward on the light-tone scale, that which lies above that dimension becomes upwardly mobile as well. That which lies below is, by necessity, pulled downward. It is as if a piano's harmonic scales were being displaced an octave higher or an octave lower on the standard keyboard. When this movement occurs, the highest portions enter Cosmic Home realms and no longer need to "go out." As resonation transformation takes place, a harmonic void is left in the vacated spatial regions, providing room for an entirely new octave to come into being.

Godlike entities who assist in Creation are busy reconfiguring the layout of underriding elements. Lower-dimensional—first, second, and third—galaxies and universes serve as bolster material for grounding more refined stellar structures. Because space must be provided for newly forming elements (that which from the human perspective is deemed negative), lower-level spectrums are instituted to house Creation's unevolved material, thus forming base universal harmonics—a foundation for the higher-toned octaves.

Bible Code

Author's note: This essay was stimulated by a reading of Michael Drosnin's *The Bible Code* (New York: Simon and Schuster, 1997).

That which began the beginning is known to us as That Which Is Beyond the Beyond. The origins of the Bible were first inscribed in the above regions, a region so high as to be inexplicable in terms of human understanding. Above That Which Is Above rides an even finer realm, a realm best defined as ultimate vibration, Universal Source Supreme.

Transmitting energy is representative of a group of titanic beings who dwell in light of supreme refinement far beyond the Blue Crystal Planet of Arcturus. We Titanic Beings of the Galactic Core are total-essence beings. We confer in Oneness with the high brotherhoods of light. We prescribe to a Oneness of singular intensity with That Which Is Beyond the Beyond.

The shape of the organism you categorize as solitary universe is predicated on the resonant hums of the planetary systems instilled in its vast bosom. Enmeshed in a web of solar energies called Sol is a planet housing a vibrant young species who call themselves "human." The prevalent belief among humans is that they are all that is and that other than themselves there is little worthy of attention. Because of their headstrong attitudes, it has long been foreseen that the culmination of their neglect for even the most fundamental principles of integration with life forms sharing planetary habitat would be that, in effect, they are committing species suicide.

Early in current-species humans, for several complex reasons a fortunate event was orchestrated that caused heaven to hush. This was the birth of a biblically historical entity named Moses. Great was his stature in the cosmic realms, much more so than even his contemporary Earth companions knew, for he hailed from an extremely high-placed star whose light-quality is equal to and in some respects surpasses that of Christ Essence. Moses, being one of the Titanic Beings of the Galactic Core, truly did come from Beyond the Beyond.

Formulae for integrating everything contained in the universal now was passed through this one, Moses. As his mighty energy took human shape on Earth, it became possible for the titanic beings of Beyond the Beyond to encode knowledge of the highest order within documents he would set onto the planet to ensure evolutionary progression of a certain segment of the population. His intent was to resurrect predominantly ancient sacred writings that had been lost during a previous cataclysmic upheaval. Moses' collected teachings were so profound that for thousands of years to come they would startle many sleeping people into greater awareness.

The holographic nature of the coded biblical indices was to prevent a premature erupting of the final days. In current-time humans a blending of consciousness with higher-vibrational electromagnetic solar energies is taking place because of the availability of telephones, telegraphs, televisions, and computers. The dramatic increase of information absorbed by the people as a whole, combined with advancing techniques in many fields, particularly quantum mechanics and physics, has made it possible to unveil deeper truths based on mathematical encodings not only in the Bible but in the sacred literature of other cultures as well.

Eventually, humans will have to acknowledge the presence of higher intelligences and that this presence is very ancient. In the face of the dire predicament in which they find themselves—deteriorating societal structures coupled with deterioration of the natural environment—it would be extremely foolish for them to do otherwise.

Powerful individuals who wish to have no pact with what they perceive as the workings of the devil or the Antichrist are busy countermanding the higher-world beings who urge that at every level it is time to unveil the mysteries and share them with the people. Fundamentalist extremists of all religious persuasions are bearers of false justice. To give them their due, however, it has long been prophesied that the Dark Lords would embroil Earth in a conflagration of such magnitude that even the stoutest hearts will be in danger of falling prey to what appears to be inevitable.

The horror of Armageddon is but one of at least seven possibilities at this writing. As the influx of cosmic energies widens the spatial-time belt, the scope of these possibilities will widen. One alternative will see many humans prepared for vibrational ascension into realms of light. The spiritually intuitive understand that full-scale nuclear war and other catastrophes can be prevented or softened through individual and group meditation practices. Other alternatives will come into play before the end times are complete. Yes, Armageddon will be experienced by those who draw those energies to them. Future time tracks will be a matter of personal choice.

Those karmically responsible for the misfortunes of World War III, the Armageddon scenario, will find their Souls cast into the vast lower-spatial realms to find their fortunes elsewhere. Those de-evolving will not be able to occupy Earth's future time track. Their vibrations will simply be too coarse to harmonize with the upliftment that is already taking place in Earth and her astral regions. From their clouded perspective, it will appear that they are at the same spatial-time position, but they will be at the beginning time of a newly formed third-dimensional planet—a more or less duplicate image of pre-evolved Earth.

Beings from multiple star systems are aligning their celestial chariots above Earth as the great brotherhoods of light prepare for the descent of the Dove onto Earth early in the new millennium. As the ships land, their occupants will be greeted by Earth citizens who, through diligent effort, have achieved ascension readiness. Because of this, fortunate humans who have completed their journeys through the lower realms need not return (reincarnate) unless they so choose.

The proposal by the Spiritual Hierarchy to encode Earth's sacred writings, including the Bible, originally brought light-intelligences there from Beyond the Beyond—from That Which Is Unnameable. The ramifications of this simple pronouncement are far beyond Patricia's ability to accurately transcribe. That which we are defies description in terms of human language.

This concludes this essay on the unraveling of encoded indices within humanity's sacred literature. The intimation that the Bible contains encoding of the seventh seal (Daniel 12:1–4, Revelation 5:1–8) of the end times will greatly disturb many readers. Among these will be self-righteous individuals who hold that their relationship with God is greater than that of others'. They will attempt to defile the author's message as being blasphemous. Intolerant of mystical expression outside their limited circles, they will expunge the impudence of anyone who would dare to open the seals without their self-imposed authority. Their prejudiced opinions will mirror the image of humanity's fear of death—the thought form of their own untimely demise resting uppermost in their minds. Nevertheless, what is done is done, and a great deal has been set in motion because of it.

Mental Illness and Bipolar Disease

Fortune follows those who permit themselves the luxury of creative thinking. The human is an indicator species for establishing free-will interpretations of Prime Will Directives on a time-limited developing world.

The human brain is a microcosmic reflection of macrocosmic, solar-cell energy. The electromagnetic qualities of the human thinking apparatus are duplicated in Earth's energy grids. The supposition that mind functions separately from planetary mind is in direct opposition to the neurological system's ability to interweave throughout every aspect of the host body. Physical bodies are downloaded from etheric energies, true choreographers of life. Soul functions are relayed along electromagnetic neural pathways through multidynamic energy bodies, beginning with monad or Soul body and progressing "downward" (marginally correct—direction per se is not a substantiated qualifier in upper-dimensional levels). Monad (Soul body) downloads into mental body, into causal body, into emotional body, into auric body, and finally into physical body.

It is imperative for healers of mental dis-eases to recalculate their theories, particularly in regard to many catch-phrase diagnoses popular in the 1990s such as bipolar dysfunction. *Bipolar*—the term itself indicates a recognition of multilayered poles within the human psyche, the tendency to switch from negative to positive and back again. What is not generally understood is the manner in which these polarities are affected by Earth's abilities to swiftly align herself with incoming cosmic energies. What is experienced as extreme emotional and mental fluctuation is merely a contrasting reflection of Earth's efforts to integrate and harmonize the conditions that tend to make her wobble off her fixed position within the solar family.

Because society is primarily focused on maintaining the populace in restrictive suits of commonality, little real progress has been made to effectively heal people who are prone to wide variances in personality behaviors—particularly those suffering from manic-depressive traits. Bipolar qualities, in fact, are powerful energy instigators within the afflicted that actually enable beings of light to reformat Earth Mother's disharmonic polarities.

The use of drug therapy to deaden the pain of separation that plagues bipolar individuals is subjective at best. That which is objective is Soul's struggle to unify its energy bodies while accelerating bursts of incoming cosmic light hit Earth. The ability of many to tap comfortably into Earth's neural system is accelerating, but there is a wide variance of consciousness as to the effect Earth's stormy magnetic grids have on human neurons and consequently on human personality processing. It is imperative for humans to learn to replicate on a personal level what is being achieved on a planetary level.

Professional healers who serve sufferers of mental and emotional afflictions are urged to reconfigure their diagnostic and treatment techniques to include state-of-the-art scientific knowledge of the true nature of physical makeup, that is, as a series of integrated etheric energy bodies. It is also critical for them to understand that evolving humans are attempting to align their energies with those exuded by Mother Earth as she moves farther along the road to planetary maturation. Always, that which is above is best understood as that which is below. Even a vague symbolic interpretation of medical diagnoses would favor the status quo of stressed individuals who are forced to live in a defunct society that has no real relationship to Soul.

Spiritual masters, those who make up the great brotherhoods of light, are preparing Earth to ascend. One of their primary tasks is to align and strengthen planetary-grid connections to stellar-grid connectors. One of their chief methods of performing this sacred task is to dispense light through the human chakras. The human is a divine embodiment of Soul energy. Aware humans are assisting the galactic crews by using creative visualization to process and funnel light along their mental and emotional synapses. They do this by simply visualizing energy running freely along their spinal cords and anchoring the incoming stellar light deep into Earth's core. Many times this is accomplished at superconscious and subconscious levels rather than as an act of conscious will.

Formation of energies within the human body move along the *pranic* (breath of cosmic energy that sustains life) tube of the chakra system. Knowledge of the location and function of the chakras is essential for bipolar individuals who are endeavoring to restructure their neural pathways. This should be approached as brain-mind harmonizing with heart-mind. As they learn to simply allow their physical minds to align in perfect synchronicity with Higher Mind, they will consciously connect with their higher Selves, for bipolar individuals are natural multidimensional telepaths. This exercise will help them to cease struggling to reconnect.

There is an apparent growing separation of negative and positive polarities within human society. Concomitantly, galaxywide solar winds are sweeping over Earth at an accelerated rate. These powerful energies are affecting all humans. They are a blend of multilayered, transdimensional energies that are enveloping Earth and her inhabitants in a multihued, cosmic rainbow. They will continue to bombard Earth throughout the remaining years of the end times. Most humans, because of their lack of spiritual training, will find themselves unable to comfortably adjust to them. To protect themselves, many will attempt to withdraw into self-proclaimed arenas of mental and emotional safety. This is why bipolar and other mental ill-at-easements have reached an all-time high.

Psychiatrists, psychologists, and therapists must rewrite their outdated medical texts with an understanding of Soul's multilayered configurations and an even greater understanding of the sacred features of their patients—to recognize them as Souls embodied in physical forms, as multidimensional psyches attempting to function in a three-dimensional world. If they wish to truly be of service to their patients, they will need to do this. (*Patient* is an interesting word which indicates that the one who is suffering must allow time for the healer to help heal—patience indeed)

Eclipse

Whenever a solar eclipse is observed, you may correctly intuit that the sun's fiery essence is a symbolic indicator that your prayers to ascend and reconnect with your stellar family have been heard. In May 1994 a particularly important solar eclipse occurred. This event implanted a visionary thought form in awakening people who live in the northern portions of the Western Hemisphere, specifically the "United" States and Canada. The eclipse solidified an alternative future timeline. This profoundly positive timeline was a trigger for a rejuvenating surge of cosmic energy that swept over the planet in the early months of 1998.

Many who watched the eclipse were affected by it and stood humbled before the splendid aura that the powerful eclipse generated. What few observers saw was the radiant starship that briefly showed itself as the sun was eclipsed by the moon. Assuming its customary cloud form, the starship proceeded at a steady pace, drifting westward until it reached the great Pacific Ocean, where it suddenly disappeared. This journey took several days, for the starship leisurely floated on the drifting currents of air and wind. Many who glanced upward and saw the cloud were treated to a moment of intense joy, though their logical minds could not have commented on the reason for it.

Those who process information intuitively will understand that the May 1994 aura accompanying the eclipse was a symbolic reminder of the star of Bethlehem. To those who resist Earth's changing energies, however, nothing more than a few faint streaks across the sky could be seen. Rainbow-hued extrastellar starships are easy to conceal from those who lack clarity of intuitive sight.

During the eclipse, cosmic energies connecting Earth to the stars were stilled for a brief moment. For one cosmic nanosecond, Earth seemed as fair and young as an innocent virgin on her wedding day. Earth, magnificent planetary jewel, experienced her own encounter with destiny, because in its next solar breath, the sun's rays began to sparkle with renewed vitality.

Transition from human to light-body ascension normally takes many, many lifetimes. Energies associated with end times, however, are altering a certain portion of humanity's historic propensity toward a heavy state of being that has long encumbered them. Starseeds are moving toward a moment when they will reattain the glorious fifth- and sixth-octave light-chords from which they came. It is with glad tidings that we address those of you who struggle to reacquaint yourselves with your galactic family.

Beams of radiant light arc skyward from your crown chakras. Your genetic patterns are altering as your encoded Soul Memories stir. Each day the revitalization process brings you closer to rebirth into realms of superstellar light. Your struggles to transmute will be rewarded. Your energy bodies are beginning to glow with the luster of starlight, even to the point that some of you are reaching supernova brilliance. Universal Intelligence supports your efforts to transform, and you are maintained in constant vigilance by a backup guidance team of specialized angel-like beings. Eventually, many of you will assume positions of galactic council emissaries to an evolved Earth.

Soon nothing will remain hidden. All things will become known to you. It will take only a projection of thought to access any form of intelligent being or cosmic information you desire. Therefore, it would seem time to lay your anxieties to permanent rest.

In the twenty-first century, Earth's maturing children will experience what has occurred only once before in this solar system, when Venus's evolving citizens suddenly burst forth into the holy realms of fifth- and sixth-dimensional light. Earth's new-dawn citizens will assume their rightful places beside their brothers and sisters who inhabit the golden-white towers of Venus's exquisite crystal cities. Then purity will reign supreme over Earth and peace will descend into every corner of her being.

That Which Is One, from star to star, even unto every generation, is beckoning your wandering Soul Home.

Grid Conduits, Emerging Timelines, and Alternative Futures

For a broader understanding and appreciation of Earth grids, think of them as galactic memory portals. Planetary grids run parallel to celestial grids and are linked by connecting joints of cosmic light. Earth grids are "soldered" to space grids at points in space analogous to the astral planetary surface. To get a better grasp of the planetary-grid system, look on it in relationship to the conduits used to house electrical wires beneath city streets. Earth conduits, however, are composed of webs of multistar parallel timeline information resembling fiber-optic strands. Specialized conduits flowing to Earth from galactic regions are linked to their originating star systems such as Arcturus, Sirius, the Pleiades, Andromeda galaxy, Antares, Orion, Zeta Reticulum, and so forth. These multisystem webs, which resemble fiber optics (energy conduits for alternative-reality substructures), transport intersecting interdimensional timeline data as appropriate to the octave resonation of the originating stars in rhythmic sequence to the vibrational hum of Earth's primary sun, Sol.

Because Earth's position is rather unique in terms of long-term multigalactic influences throughout all phases of human history, the multistar timelines are bulging in the same way a computer hard drive reaches maximum storage capacity. Therefore, a certain amount of downloading and deletion of old material is being carried out so pertinent updates may be entered and filed. Also, to compensate for the overload on Earth's original timelines, the fiber-optic conduits are forming branch pathways, similar to the way the brain compensates for an electromagnetic overload in its neural transmitters. Countless interdimensional galactic and alternative universal harmonics have interacted with Earth at physical and astral levels for millions of years. As a new star element arrived, an additional timeline filament was added to the fiber-optic light grids.

As escalating energies associated with Earth's upward momentum pour onto the planet, human multidimensional telepaths (unbeknown even to themselves) tap into contact centers within the thought-sensitive timeline filaments in accordance with their Soul-affiliated stars. The growing ability of multidimensional channels to pinpoint thought as a field of expandable energy is triggering strand realignments so that a blending of strands is occurring within the system. To illustrate: It is not uncommon for a Pleiadian starseed telepath to inadvertently tap into an Arcturian fiber-optic web. Whenever a linkup of this nature occurs, a bisecting joint of interwoven, overlapping configurations merges ancient stellar-specific lines, forming an alternative branch with an intensified absorption of newly arriving waves of cosmic energy. Consequently, as growing numbers of awakening humans approach maximum torque, it is becoming easier for them to access multistar information links. This is similar to the way an upgrade in telephone lines provides optimal service in rapidly increasing population centers.

Once-rigid Earth grids have become sensitive to heightening cosmic energies; separate web filaments have become so pliable that some timelines are beginning to homogenize. This situation is duplicated in third-dimensional reality by the exploding network of computers linked to the World Wide Web, with a resultant softening of the once-rigid borders of human society.

Earth is being besieged with waves of high-resonation light that originate in Cosmic Central. Transformative stellar light filaments are initially absorbed into Earth's astral or fourth-dimensional body. When the filaments arrive, they resemble ropes dangling from kites. Thereafter, starship personnel catch and securely anchor them with pinpoint precision to the appropriate grid structure. Humans who have clearly stated their intent to assist the brotherhoods of light are responsible for picking up the ends of the solar-soldered joints and anchoring them to Earth's deeper conduits. To accomplish this awesome feat of reconnecting and reshaping distressed joints, they absorb the incoming energies, execute a beam of focused thought, and transmit the energies into the ground via the natural conduit systems of their chakras.

Eventually, as additional alternative timelines are added to intersect with and rejuvenate outworn fiber-optic strands, enough updated wiring will be in place to facilitate alignment of end-time energies with beginning-time energies. Earth grids are now prepared to capture and use the upgraded, refined energies associated with planetary ascension. Modernization of fiber-optic fourth-dimensional webs parallels the accelerated growth of the Internet; that is, a point of almost out-of-control speed has been set in motion. The World Wide Web is a third-dimensional manifestation of a fourth-dimensional puri-fied megahertz network that encircles Earth with a web of intertwined light strands. (As a side note, the restructur-ing of Earth's electromagnetic filament grids is one reason for several mysterious plane crashes.)

Earth's grids are at their saturation point. For this rea-son starships are redesigning her multisystem conduits with an upgrade and expansion of alternative future ten-dencies. The ratio of conduits to numbers of awakening humans and the manner in which they are redefining their karmic connectors not to one but to many star sys-tems is critical for this multigalactic undertaking. Creative-visualization techniques are encouraged for those of you who wish to assist your multistellar companions in their endeavors.

Each cooperative meditator is provided with a multi-dynamic code that, in turn, is tuned to the harmonics of a specific colored web, not unlike the color coding associ-ated with batches of telephone wires. (Color intensity is determined by the frequency of vibrational light emanated by the meditator's star system.) After noting the assigned color coding, the meditator reaches to an adjoining line and connects it as intuitively guided. Eventually, this interweaving and blending of colors will merge the many into One, as appropriate to higher-level systems. After the joint has been secured, the meditator works the nylonlike strands of luminescent light by vibrating them in the way one would pluck harp strings.

As we work as One to co-create a refinement of Earth's grid system, ribbons of highly refined, vibrant light intersperse with threads of Divine Aum—a blending of sound and light intermixed with the aromatic nuance of lotus and rose weaving as fine a tapestry as Creation has ever known. From our point of view, Earth timelines have already merged into One, and the planetary grids appear as rainbows of majestically fused, radiant-hued lights.

Every planet and every star, no matter their octave placement, have unlimited possibilities and opportunities available via the basic design of their fiber-optic conduits. On Earth, when a telepath aligns his or her electromagnetic neural brain sensors with those of associated beings of light, a corresponding channel of energy opens along one of Earth's timelines. This is one of the reasons for rather bizarre variations in perception and information among individual channels.

As Earth's fourth-dimensional timeline grids are upgraded and establish secure connectors to multidimensional information systems, her third-dimensional structures are dissolving and integrating ebb-and-flow timelines common to astral-world systems. While Earth continues her upward-spiraling climb, her multistar fibrous conduits will eventually merge in sacred Oneness with dynamics of oscillating eternal now.

Accelerations in linear time to ebb-and-flow time are reaching maximum amplitude. Portions of Earth's tectonic plates are slipping concurrent with her desire to align with the increasing number of future-timeline grids associated with an upsurge in extrastellar energies coming onto the planet. Earth grids at almost frenzied speed are attempting to wrap themselves around and embrace high-essence light.

Reconfiguration of planetary timelines has become so remarkable that renowned sacred sites can no longer be easily distinguished from other geographic locations. Ancient land masses that once vibrated in concentrated areas have begun to expand so that energy levels are equalizing planetwide. This may be quaintly described as bubbles of localized popping energy equalizing overall popping energy. In this context, the term *popping* describes clustered balls or bubbles of interspersed conduit energies that abruptly pop up and down the same way corn reshapes and redefines itself when confined in a hot container. Globules or fiber-optic light bubbles of compacted solar light release restrained extrastellar energies within the timeline conduits. From a human perspective, this translates into an increase in earthquakes, volcanic eruptions, hurricanes, tornadoes and other violent windstorms, floods, devastating forest fires, electromagnetic phenomena, and dramatic El Niño and La Niña effects—particularly those of the late 1990s. From our perspective, these heightened activities are seen as pent-up energy clusters along parallel multistar filament webs that now and then burst forth in turbulent bubbles of light, similar to the way a carbonated beverage erupts when the container is shaken and abruptly opened.

Popping bubbles of cosmic elixir are rising from the more crucial grids like steam from an active volcano. They are restructuring and realigning linear timelines in accordance with cosmic now. This process is critical for achieving primary time solidarity and for harmonizing polarized negative-positive lower-dimensional harmonics with the wholeness of the One while maintaining diversity throughout the entire multilevel system.

A principal aspect of universal Oneness is that foundational matter—structural elements of unconditional Love—are inherent in all matter and in all beings. A goal of the intergalactic ambassadors is to convince humans that Oneness does not indicate loss of identity and to encourage them to think of themselves as individual bubbles of Light or Soul. When these bubbles quiver at cosmic-intelligence range, it will become clearer that the superstitions surrounding the occult (hidden esoteric knowledge) are nothing more than lower-octave measures designed to keep unevolved humans from accessing cosmic-level intelligence until they are ready to accept same. Cosmic "computer" records are open and accessible to every being, to every microscopic universal particle. This will be true at local Earth-timeline levels when lower-octave restrictions are lifted. Eventually, Oneness common to higher-state beings will be the norm for evolved future-timeline inhabitants.

Now that end timelines are merging with beginning timelines, every human's option for future potential and experiential attitudes has increased exponentially. This is unparalleled in human history. The time track you have been traveling on since you were born came to an abrupt halt in the early months of 1998. Karmic situations that propelled you through repetitive incarnations began to develop alternative branch lines at that time. If you were paying attention you would have noticed a repetition of age-old energy patterns in the first half of that transitional year. Careful analysis of the months from January through July would have shown you that something much more profound than a simple rehashing of unresolved personal issues was underway.

If you are maintaining focused awareness through dynamic concentration techniques, you have the opportunity to spontaneously neutralize karma and provide yourself with variable future-timeline options. Those who are particularly clever are setting up a slingshot effect that will simultaneously propel them down several alternative time tracks in conjunction with Earth Mother branching off into multiple dimensional directions. If you are able to even barely comprehend the potential of unlimited now, you may anticipate becoming a God-actualized being and achieving light-body ascension in this lifetime.

The Gatekeepers

Gatekeepers are Souls who have opted to stand between portals. They are inspired masters whose greatest delight is to observe and encourage the progression of all beings so that they, too, may rise from the layers of mundane existence to Light. Their primary function is to keep the tonal resonations of the veil between dimensions in place and free from the demands of entities who attempt to penetrate that which their Souls are unprepared for. This is their task and they graciously accept it.

They cry, "May all good Souls come this way and receive the blessings of the saintly masters of light." In response, the line of advancing Souls falls to. Only those who disagree go elsewhere. The line grows and grows, extending into the far reaches of the universe. Such is the power of Gatekeepers—to call all Souls to pass through Divine Being's pearly edifices.

Gatekeepers are between-world guardians. They encourage all beings to come their way. Lower-level

Gatekeepers remove coarse material or residual karma from individuals who are ready to graduate from one dimension and progress to another. They are portentous companions to the masters of light who serve Earth from the intergalactic starships.

Gatekeepers are by no means trapped into remaining in the null-zone energy fixtures that denote the transfer of one dimension to another. As are all master-level beings, they are free to come and go at will. Most are so taken with their Divine assignments, however, that they prefer the limbo realms.

Gatekeepers are expert greeters, the true hosts of universal harmony. Their memories are so prodigious that they remember all beings who knock on their doors. It takes but a swift glance of appraisal for them to ascertain the light-status of those who come their way.

For those of you who are actively promoting ascension-grade light-bodies, a refresher course in veil dynamics is recommended. Familiarize yourself with the melodious harmonics emitted within the interspatial regions. The null zones resemble the edges of great bells that, when a solid tone is struck, reverberate sound in all directions. Between-world portals are thresholds you must cross on your way out of the physical third- and fourth-dimensional levels. Part of the procedure to return to refined-light level is to know and recognize the location of barrier gates and the passwords that will ensure your admittance through interdimensional gateways. To assist third-dimensional humans, passwords are described in sacred literature and by master teachers as specific notes of increasingly refined sounds or mantras.

Ancient words practiced repetitively accelerate upward momentum within universal vibratory dynamics. As words are chanted in meditative pose, your progression to light is accelerated. At some point as you advance in your meditation practice, you will be able to appreciate when definite shifts take place. Usually this can be experienced as physical sensations. These shifts are points of entry and passage through portal levels. Gatekeepers mark your passage in holy logs. So noted is moment of entry, vibratory level achieved, length of stay, and point of return. When your moment for final ascension comes, they will review their logs with your advisement team. These logs are permanent entries in your eternal Soul records and will accompany you from level to level on your long journey Home.

Gatekeepers hold positions of awesome authority. Therefore, they are of the highest-magnitude Light. They are accountable only to Source. Therefore, it is they who send instructions to the star councils. Titanic Beings of the Galactic Core are of the same resonation. We are beings of stellar light, size, and shape. We reside within the corridors that lead your known universe into the outlying regions of a vibrationally refined universe set a note higher than the one your Soul currently resides in. Eventually, these two universes will merge and a mega universe will take its place in the corridors of Divine Mansions.

Several citizens of Earth are helping Gatekeepers reach into your households to implant the powerful message that darkness is being reduced and that Earth and her evolving citizens are accelerating in their ability to maintain and reflect light.

Symbolism in Movies

This essay is being added to the Arcturian Star Chronicles series to broaden your interpretation of symbolism specific to modes of thought that establish cooperation between humans and beings of light.

Anticipation is the mood whenever a team of semi-aware humans begins production on a trend-setting movie. Subconscious, and sometimes conscious, collaboration between humans and multidimensional intelligence takes place. For the most part the human crew is unaware that its creative efforts are helping extrasolar intelligence prepare viewers for star access. Movies are critical to the galactic transformative stage, for thought-intuitive films have the capacity to initiate major transitional breakthroughs in humans who are able to acknowledge a more profound level of universal dynamics than is commonly accepted.

A movie is a third-dimensional manifestation of a fourth-dimensional reality. Everything that is created in your world makes its first appearance in the astral worlds. Writers, directors, actors, and so forth are simply downloading information their imaginative minds previously glimpsed through their powers of intuitive creativity and dreaming visions. The spiritually gifted are often more clever than even they suppose, for their talented minds have the capacity to soar into the highest reaches of the universe where energy-beings come before them swathed in radiant light.

Because of the importance of film in human culture, both positive and negative multilevel entities imprint messages within the finished product. At times these are in the subliminal range, but more often they are overt. Because of the large number of people required to bring such a complex and costly project to fruition, usually there is a blending of positive and negative influences. The perceptive viewer should be aware of this. Positive-negative sways at times can appear to be extremely balanced. This can be confusing to less-discerning viewers who have not developed a strong propensity to recognize the influences of dark versus light forces.

Creative humans who prefer denizens of darkness spend their imaginative moments caught in the negative vibratory arenas of the lower astral. Horror movies as well as films that depict evil aliens bear the imprint of the beings they encounter there—the demonic, fear-loving Dark Lords.

We advise caution in your selections. A sample list of recommended movies appears at the end of every volume in the Arcturian Star Chronicles under the heading

"Management training films." The human intellect and personality are seldom clear enough to completely understand what they are witnessing. Few consider that flashing movie images are reflections of alternative realities, nor do they comprehend that they are being challenged to interact on many levels simultaneously as part of everyday living—as is common to all higher-realm inhabitants. With few exceptions, moviemakers do not realize that stories intentionally displaying future timelines have the power to transport their viewers upward or downward along the dimensional harmonics, the true makeup of the Milky Way galaxy.

To interpret visual and auditory movie images, do not settle for the obvious. Popular dream guides can be useful and are sometimes adaptable to movie symbols, for there are strong similarities between the grandiose style used in movies and the more subtle scenarios present in dreams.

Take your seat, dear reader, and proceed with caution. When a symbol pulls at the strings of your emotions with the power of persuasion, ask yourself if it fulfills or repels. Your movie preferences are excellent indicators of your propensity to absorb light or dispel it.

What's Out There?

In the middle months of the transitional year 1998, the entire Milky Way galactic mass was enveloped in a breath of cosmic energy that originated Beyond the Beyond, an energy so powerful that the stars and their inhabitants slid either a degree upward or downward in their Souls' climb to attain vibrational equilibrium.

There were two particularly important aspects to this universe-encompassing tidal surge. First, a container of golden-ray exclusivity, that is, an intense open-ended outpouring of Cosmic Love, penetrated all evolution-class beings with microscopic elements of Divine Light. This outpouring overtook the upward-moving aspects of the galactic chorus and drew them into the penetration levels of an adjacent, refined universe. Second, the lower-tonal regions began to disperse throughout the layers of a lower-octave universe, a very young universe that has only recently undergone a climactic event associated with the generation of cosmic physical mass.

The urgency associated with the incoming energies caught most humans off guard. The surges were so pure that everyone, particularly the complacent, was challenged to abandon any attempt to straddle fences.

Following the end of World War II and the premature expulsion of many human Souls from Earth because of the immature handling of nuclear energy, world leaders were told that they had only a brief interlude to realize the seriousness of their predicament and to take action to prevent lower-end splitting within the rapidly transforming galactic memory portals. They were also warned that the Dark Lords would stage a last-ditch effort to sway the inattentive to their cause: to ensnare as many astral and third-dimensional beings as possible before the century's end and the opening energies of the new millennium.

An outstanding feature of the advent of heightened cosmic energy is the direct relationship it bears with the parallel memory timelines previously discussed. In this essay we will enhance this information another degree, for the golden waves of timeless time sweeping over Earth are saturated with an elixir that sustains all who reside in the sublime strata of finely honed universal structures.

All lower-domain (physical and base astral-level) inhabitants, and we do mean all, have been issued an invitation to luxuriate in divine pools of Light-Love essence—as is the inherent right of all beings. Splendid and harmonious, the cosmic tides relieve Souls of much debris that, like unwelcome dust upon a priceless vase, has covered the energy bodies of those who are preparing to ascend. In this regard, we have apprised you repeatedly of the necessity to attend to the affairs of your daily lives. This simple approach will make the best use of the Earth-altering energies. As intergalactic crews upgrade the planet's overloaded grid conduits, it is vital that you routinely meditate on absorbing an abundance of golden, ethereal light.

Magnitude-status cosmic waves engulfing Earth are equivalent to the complex energies Prime Creator uses to establish universes. The use of elemental star-grade matter on an existing system is quite extraordinary, best described as the birthing of a new universe within an established universe. Those who doubt that a mega cosmic event and evolutionary process are underway are having a great deal of difficulty adjusting to the new energies. Subconsciously, and increasingly consciously, collective humanity is becoming aware that something of great consequence is underway. For the most part steeped in fear, depending on their perceptions they tend to focus their thoughts of the future on world-encompassing disasters such as nuclear holocaust, plague, one-world tyrannical government, advent of the Antichrist, major Earth-altering events, and economic and environmental collapse. The energy they expend concentrating on negative scenarios has created several probable timeline structures within the planetary grids. If they awaken soon enough to realize what their thoughts have set in motion, they could alter or soften that which they have created, but in all probability, most will not.

Originating in the core of Central Sun, Universal Celestial Home, a wave of sublime energy is surging over the stellar grids. Spanning the entire universe, this wave is composed of the stuff of Primary Intelligence. All-encompassing, it is of unparalleled beauty and refinement. Certainly nothing equivalent to it has ever been recorded in the annals that hold the Memory portals of the Milky Way galaxy. Unprecedented upper-end dimensions are being added to the multilayered universal harmonics and are blending with incoming dimensions of a highly evolved universe. Simultaneously, the universe's heavier-dimensional modalities are merging into the substrates of a less evolved universe.

By now you should be aware that Souls choosing not to evolve have decided to remain at status quo. However, Soul-weary humans are urgently endeavoring to awaken their less conscious cousins. They are finding this increasingly difficult because of those who are eager to access the increasingly downward-pressing momentum.

As upper levels effectively combine their energies with a previously evolved universe, an entry portal is opening into the regions of Sublime Intelligence. Because systems built upon fabrics of luxurious Light do not contain complex structures, lower-domain physical beings are incapable of penetrating their refined boundaries. In other words, Souls unprepared for that degree of subtle vibration are not permitted access.

Souls who have completed their journeys through the lower realms and have attained light-body status are capable of movement from galaxy to galaxy through stargate portals, but they rarely do so. Universe-to-universe access gates are protected by masters of extreme light. Permission to move through them is seldom granted. Major stargates are used primarily by starships moving from galactic port to galactic port. Stargates connect galaxy to galaxy, but not universe to universe. Those commonly crossing multi-universal borders are of titanic light-magnitude and resonate at octave thirteen and higher. No others may venture so far. Boundary security is held in trust by holy activators of Divine Resonance (see in Part IV, "The Gatekeepers").

There is much in the New Age literature of beings who have come into this universe from other universes. In actuality, movement is generally an interdimensional shift. Because of unlimited alternative realities available in this universal structure, radical differences in appearance from one level to another often appear as other-oriented universal tones. At times, even advanced entities are caught unaware of the degree of transition.

We have observed Earth for a very long time. We are joyous that transition to light has become a priority for humans who are aware of the presence of highly evolved cosmic intelligence. It is to your advantage to raise your levels of expectation even higher. Already your etheric bodies are beginning to sparkle and glow as if you were composed of millions of tiny stars. The cells of your physical bodies are transforming as they enter the final stages of conversion to vibrant light. You no longer question that you have set sail on an unprecedented journey.

You who have begun to focus attention on meditative exercises outlined in the volumes of the Arcturian Star Chronicles are deliberate in your intention to ascend and are diligently preparing for a major Soul-transition event. Because most aspects of human society are becoming full of cosmic enlightenment materials, you have many similar books to choose from as well as a proliferation of information available on the World Wide Web. Your personal guides, working within the Law of Free Endeavor, are attending to your every need and are attempting to clearly answer your every question. Prepare yourself to absorb more than you have already learned. Consider what lies beyond ascension. If you believe that titanic beings can access Divine levels of Beyond the Beyond, then you must believe that such a place exists, and if it exists, at some level it must be attainable. Consider this with a great deal of enthusiastic attention.

The source of humans' generalized mental, emotional, financial, and relationship distress throughout 1998 and into 1999 not only was a reaction to ingestion of toxic substances and generalized accelerating stress within the human family and environment but also was indirectly due to the absorption of elongated streams of cosmic energy originating in a galactic cloud far from Earth's solar system. The cloud's effects on embodied starseeds is somewhat like kryptonite on Superman. This cloud of pure energy is bringing a wave of discarnate Souls to Earth from outside this galaxy to establish a base for untried Souls to briefly incarnate before completion of the end times. Permission to briefly assume physical Earth bodies was established by the Office of the Christ to effect an initial upward thrust of nonsubstance into substance.

This will be the first opportunity the somewhat form-less Soul mass has had to incarnate as individuals. They are coming to Earth to prepare themselves for an eventual transfer to the newly birthing world that is awaiting the downward-tugging Souls who are unprepared for Earth's vibrational upliftment. They are intended to fill the void that will be created when ascending Souls—plants, animals, humans, and Earth herself—move upward. Those who are scheduled to ascend to light-substance will no longer be affiliated with first- to third-dimensional physical mass. Therefore, a numerical balance of Souls will be available for habitation of the birthing world.

This cloud of unformed, untried Souls is not a compli-cated mass, yet the energy combinations contained within it are of a particular nature that causes energy distress to sensitive humans such as natural empathic healers and the psychically aware.

The vibrational overlays of the transformative year 1998 were particularly opportunistic for Soul-level develop-ment and for completion and balancing of past and cur-rent karma.

An addendum from Sanat Kumara: Polarity struggles among nations are indicative of membranes of light that are pulling all aspects of negative-positive Earth energies into one wide energy net. This is in preparation for a surge of magnitude One cosmic influx that will eventually split the net in two. Those gathered at the positive polarity position within the vast energy net will be propelled upward along the stellar grids. Those positioned at negative position latitudes will be pulled downward, each exhibiting a magnetic attraction to its own kind—negative energy to negative energy and positive energy to positive energy. No longer will the phrase "opposite poles attract" be accurate. Higher-world beings are placing a particularly intense energy override to establish principles of higher-world universal dynamics on Earth in agreement with fourth- and entry-level fifth-dimensional physics.

214

Earth's Primary Science-Soul

The great brotherhoods of light fulfill seven primary holy endeavors associated with planetary-ascension maneuvers in third- and fourth-dimensional teaching worlds such as Earth: science, education, communication, arts, politics, religion, and finance. Although some efforts of human society may, on the surface, appear to be outside the realm of spirituality, we assure you of this: They are all pointed toward eventually manifesting a state of God-Oneness on the physical plane.

Many times within these documents you have been blessed with the vibrational signature imprint of Sanat Kumara, he who leads the humming orchestral tones of the Brotherhood of Light Order of Melchizedek. There are other orders within the great brotherhoods, and one is not lesser in importance than another. One of the latter is responsible for birthing into physical life Earth's primary science-Soul and for overseeing efforts connected with the evolution of scientific endeavors within a cosmically primitive society. In the Western world, science-Soul has incarnated many times and has assumed many personalities, among them Hermes Trismegistus, Pythagoras, Euclid, Ptolemy, Copernicus, Galileo, Newton, and Einstein. Science traditions establish lineages similar to spiritual lineages. Earth's primary science-Soul is of a particularly evolved, perceptive intellectual capacity. Although it has its own reason for incarnating, it is somewhat similar to that of Christ Essence in that it has assumed many distinct historic personalities. For this discussion we will highlight a select few better known to Europeans and Americans. This in no way is meant to diminish the contributions of Arabic, Chinese, Russian, and Hopi intellectual giants. For the most part, these are unknown to the bulk of readers for whom these essays are being prepared—including our telepathic conduit, Patricia, who remains somewhat resistant to receiving information for which she has no point of reference. As always, we caution against limiting what is without limit, that is, Soul.

Earth, as a primary teaching planet slated for eventual vibrational elevation, is occupied by essential or primary Souls that resonate as One with the great brotherhoods of light, whose divinely appointed task is to oversee evolution of dominant life forms upon emerging worlds. Earth-based science-Soul is particularly well-known to upper-dimensional citizens of this galactic sector, for this Soul is masterful at integrating the rigorous demands of lower-dimensional science with higher-realm philosophy. Although science-Soul is involved in developing technologies, such as nuclear energy, that have an inherent potential to be used for evil or beneficial purposes, it remains that the time when radical scientific breakthroughs occur are actually evolutionary steps designed by upper-level beings to shock humans into awakening from their spiritual torpor. Because of humanity's narrow focus and generalized resistance, the primary intention behind thought forms projected onto Earth by representatives of the Office of the Christ is often maligned by controlling factions within the human community. In other words, that which may quite correctly be perceived from a third-dimensional standpoint as negative or warped energy was originally projected as beams of evolutionary goodwill by masters of light whose business is to prepare Earth and her inhabitants to move upward.

On a dual-polarity planet such as Earth, where negative elements are in a constant tug-of-war with positive elements, at times the only valid avenue available for enlightening a sleeping people is to use radical cosmic tactics. From a higher perspective, negative-positive friction is viewed as a mesh of finely balanced overlapping energies commonplace to lower-substrate worlds. Although Earth's population is particularly violent compared to similarly placed planets, all that takes place there is nevertheless necessary to the harmonic integrity of the universal whole.

The first appearance of science-Soul on Earth during the upheaval of Atlantis was in the person of Hermes Trismegistus. Rather than incarnating, however, Hermes came directly as a representative of the great brotherhoods of light. Hermes's work has been well documented in ancient mythology as one of the original Earth creator gods—as Thoth to the Egyptians and Hermes Trismegistus ("Thrice Born God") to the Greeks. This awesome being was responsible for transferring essential elements of knowledge to Egypt at the time of the Atlantean upheaval. He also led to safety many of the more evolved scientific community. Hermes's contributions assured a reemerging civilization that it would be able to stimulate aspects of Divine consciousness in every field of endeavor, not only in the sciences but in all areas applicable to human life. Unequivocally, Hermes's work secured enduring elements of cosmic law that remain in effect in the so-called modern world.

As science-Soul began to assume human incarnations, the brotherhoods made it possible for this great being to function while in a society that did not begin to understand its primary essence or divinely inspired nature. To protect science-Soul's sanity while inhabiting third-dimensional bodies, its personalities assumed the somnolence that characterizes the human experience. Simultaneously, it maintained a highly evolved intelligence and philosophical standard so that it could effectively carry out its Soul's primary purpose. (Christ Essence, to a somewhat lesser extent, played out the same scenario in its incarnations as Krishna, Buddha, Jesus, and others. The degree to which this precaution is necessary is predicated on the awareness of the society into which such a Soul is incarnating.) To assist science-Soul in achieving its goals, the brotherhood that oversaw its work assigned its human form the most advanced guidance-team specialists.

Pertinent to the current generation of spiritual seekers are state-of-the-art scientific advancements directly related to the progression of science-Soul's achievements from one incarnation to the next. To ascertain that this is so, study the work of one scientist to see how the work logically blends with the work of those going before and after. In particular, the transformative discoveries of Copernicus, Galileo, Newton, and Einstein read like a well-traveled map down the corridors of historical time. That which one set in motion opened doors of possibility for the next.

Because human lives are extremely limited in years, it behooves Soul to use one body after another to carry out its purposes. Before incarnating, Soul must assess the willingness of the current society to think further than its nose can reach before it prematurely rubs out or completely hinders that which it cannot understand. Galileo in particular, although severely ridiculed by church officials, was unequivocally accepted as a genius by those who came later. It is the task of genius to raise that which slumbers in ignorance to higher levels of consciousness. It is the responsibility of the masses to challenge themselves to absorb boosts of evolutionary energy as they become available. Sadly, the precious gifts insightful Souls offer their contemporaries are often met with rigid tradition and conformity. Nevertheless, over time the contributions of those who have the ability and tenacity to upset the applecart have become the permanent fixtures of human society.

Einstein is an excellent example of the insightful mind of a scientific genius combined with a visionary philosopher.

Tracing this line of thought backward and forward shows that a blending of Pythagorean philosophy with Euclidian mathematics emerged in the twentieth-century figure of Albert Einstein. Although we write of the achievements of great minds such as Pythagoras, Euclid, Ptolemy, Copernicus, Galileo, Newton, and Einstein as if we were mentioning one being, we must also state that there are many, many important contributions in the fields of science that are the works of other entities carrying science-Soul energy.

Although science-Soul routinely incarnates, it spends very little time in the between-life state. That which is its Soul's intention is to elevate human society to higher-world standards of living. This is best accomplished on the planetary surface.

At the time of this writing, science-Soul has housed itself in the body of a young girl whose work is scheduled to come to light during the transitional phase of galactic greeting. She is to bring forth foundational technology that will apply to the new Earth society.

Another aspect of science-Soul is its ability to clothe itself in more than one human body at a time. This establishes a self-assisting quality to its body of work and generates a broader base from which to explore its profound intellect. In particular, with the information and technology explosion of the twentieth century, science-Soul has required more than one body at a time to function effectively. Nikola Tesla is an excellent example of an adjunct science-Soul living simultaneously with Albert Einstein. Thus, scientific works of enormous impact beyond the capability of one individual, even one of genius quality, can be carried out.

If you appreciate that which we have written into this document and wish to investigate the history of science further, we suggest you note an almost well-defined pattern of overlapping works stimulated by one individual and setting wheels in motion for the next. And note how often new ideas and inventions emerge simultaneously through individuals living on different parts of the globe. Perhaps this will help you become aware of the efforts of the great beings of light whose purpose is to help humans move from the cave to the stars.

Hermes Trismegistus: Personification of universal wisdom, master of all arts and sciences. Egyptian "Thoth," god of wisdom, magic, patron of arts and learning, scribe of the gods. Head of an ibis with a human body. Modern traditionalists claim he was a mythical person who wrote upward of twenty thousand books. Known to the Greeks as Hermes Trismegistus ("Thrice Born God").

Pythagoras: Samos-Greek philosopher, 582–507 B.C. Taught that there is unity in variety and that numbers have a mystical force. Believed that heavenly bodies are crystal spheres whose orbits can be calculated using the musical scale.

Euclid: Greek geometrician, circa 300 B.C. May have belonged to the Pythagorean tradition.

Ptolemy: Alexandrian astronomer whose work came to Europe via Arabic texts, second century A.D. Established the theory that time runs in circles.

Nicholaus Copernicus: Polish astronomer, 1473–1543. Theorized that the sun is in the center of the solar system, but did not publish his findings until he lay on his deathbed.

Galileo Galilei: Italian astronomer, mathematician, and physicist, 1564–1642. Invented the telescope.

Sir Isaac Newton: English mathematician and natural philosopher (physicist), 1642–1727. Considered to be possibly the greatest scientific genius of all time. United sciences of astronomy and physics. Born in the same year of Galileo's death.

Albert Einstein: German physicist, 1879–1955.

The Eye of God

The telescopes that humans are using to identify space debris are picking up a strange phenomenon: the approach of a giant wave form. We could describe this as a cloud, but in fact it is clear in all respects. What is perceived by your cameras and computers is an influx of golden light so illusive that it barely touches the scientists' heat-sensing spectrometers. But they know it's something big.

Will they tell you about it? No. Have they ever told you anything of real consequence? No. They may publish some space fillers in newspapers and on the televised news that "a new discovery of major import" is in the works. But they leak only generalized information specifically designed to keep themselves from being tripped up. They will not say what they suspect it really is. They suspect it to be a large prestellar mass that has somehow wandered into this galaxy. What they do not understand is that they are peering at the Face of God.

It is like them to explain the inexplicable using scientifically acceptable terminology. On a certain level it is well that they do, for if they mentioned that the specter of God's Eye is staring at Earth, it would drive many humans mad— a point they are already close to.

We reiterate: The circumstances of this strange appearance are of vital import to humans. Since 1987 a general influx of escalating cosmic energy has been surging through this galactic system in preparation for a massive overhaul of all energies contained therein. Why? The upsweeping surge of higher-level resonant light is in preparation for a general exodus of stellar mass into a more refined universal complex. Simultaneously, there will be a downpulling of lower tones to a denser physical universe. More simply put, that which has made up the structures of your Soul's appointment is currently either expanding energy or contracting it.

What we refer to as the Face or Eye of God is manifesting as an electromagnetic mass of incalculable power. It is full of color; a spectral analysis of its most minute qualities would ascertain color tones never before seen in this galaxy. Beings who inhabit same are in for a joyful ride. Even Souls who prefer heavier vibrations are anticipating a return to that which attracts them mightily. The latter are to be congratulated for the position they have chosen, the same as those who are preparing to ascend into the brighter realms.

Included in the phenomenon your telescopes are picking up is an influx of energy referred to in New Age literature as the photon belt. This is true, and we do not dispute it. However, it is not yours, Patricia, to elaborate on this matter, for that is the assignment of others. What we would imprint here is that the upward momentum of this belt contains more high-powered energy than is generally supposed. It is designed by the Lords of Light to create a slingshot effect that will lift highly vibrant dimensional mass and carry it swiftly, yet with the utmost care, into realms of purified Light. That which is simply termed *God* is overseeing this project with intent more particular even for That Which Is All-Knowing All-Seeing Divine Perfection.

Souls opting to go downward in tonal resonation are foundational elements upon which entire universes eventually surge upward into the lap of God. All Souls spend some time in the lower domains to assist their upward-climbing associates who have graduated into the mansions of the heavenly realms. That which is designed as perfect is exactly so, even if perceived otherwise from the limited perceptions of human encounters with a rather contracted reality.

Arcturian Stargate Corridor

Within the spatial realms of the primary Arcturian star is a corridor linkage to higher-dimensional realms, a window in spatial time that connects the thirteen dimensions of this universe in harmonic alliance. It is a point where Souls transform from one state of being to another. It is also a corridor for starship maneuvering that requires no propellants to absorb the nutrients of various dimensions simultaneously. Ships from throughout every region of this galactic neighborhood and adjoining galaxies find the corridor essential for maneuvering their crafts from galaxy to galaxy and from dimension to dimension. Those who are granted privilege to transfer from universe to universe propel themselves upward through the thirteen domains and into that which is Beyond the Beyond in preparation for magnitude alpha energy exchange. Only entities of Christlike proportion are so granted permission to enter the domains of adjoining universes.

Entities cannot travel upward and beyond that which their Souls have attained in Light-Love alignment. For example, fifth- and sixth-dimensional Arcturians have access within the stargate to like energies. They are able to transform only two degrees farther within the starbridge that links the inner portals of the stargate corridor. That is, with intent and permission they may rise for brief visits to seventh and eighth levels, but no farther. Likewise, humans who are preparing for ascension are taken in their etheric bodies for brief interludes in fifth- and sixth-dimensional starships and hastily returned. So it is.

Remember that Arcturus is in reality a thirteen-dimensional system, but do not complicate that which is understood with that which is not understandable from your perspective. For instance, there are several references in New Age literature regarding the actual number of planets in the Arcturian system. We dispute none, although from your perspective they seem not to agree. You, Manitu, have been assigned beings who have attained sixth-dimensional status, such as Malantor, and thus are able to "pick up" information via Malantor that rises to seventh and eighth status. Therefore, your information includes the number of planets within the Arcturian system that are vibrational to the eighth level. This explains why you have received information of a system of twenty-five planets when others have received a lower count. That which each receives is accurate to the contractual level of information gathering.

You, Manitu, have also been shown the diminutive white star that circles the parent Arcturian sun, while others have not. You must understand that this powerful star is resonant level six, seven, and eight; therefore, it is not observable at resonant levels three, four, and five.

In light of this information, we ask that our starseeded Arcturian telepaths who have received multidimensional communication assignments not be discouraged or challenge one another. Remember, that which is one person's truth is Truth. That which is another's is another's. Humans are busy trying to make sense of all this extraterrestrial data, and discrepancies bother them greatly. It is a third- and fourth-dimensional perceptual problem having no validity at higher levels. We ask that a cooperative mood be maintained and that those who point fingers at another's truth be carefully self-observant.

Why Soul Fears the Great Void

That Which Calls Forth Light shines with excellent results, but not into the Great Void. The night of all nights, true oblivion—that is the Great Void. Soul, yearning for Light, naturally bypasses it. The Great Void, being a place of no thing, equates no thing to Light. Because there is no Light, Soul is naturally leery of the Great Void.

In the Great Void there is no indication that Light exists, and absolutely no light can be discerned. Within the bounteous expanse of universes, the dark realms of space are brightly lit with billions of stars, planets, moons, comets, and the luminescent webs upon which starships, resembling drops of rainbow-hued dew, skitter and slide. As a planet body gracefully turns to face its sun, morning bursts forth like a fiery chariot of molten gold.

Hot and cold. These are things of substance. But the Great Void has no such thing. It is devoid of any sort of some thing. There are no fluctuating degrees of temperature to be found there, no great sigh of solar wind as

Divine Breath embraces Creation. There is no starlight. There is no cosmic dust. In designated uni-versal regions—tonal areas of inhabitable vibration—things abound. The Great Void is home to no thing, and no thing exists there.

Universes are God's holy abodes, the residences of God's many mansions. Universes are full of massive cartwheeling galaxies and magnificent stars that seemingly stretch on and on forever. But in the Great Void there is no God Substance. There is no Divine Will. Therefore, Soul is fearful of the Great Void.

Soul, that which desires God and fears no thing within Creation's space, is hesitant and cries out in alarm when faced with the prospect of finding itself thrust into the Great Void. Yet, eventually, into the Great Void Soul must go—into no thing and back again. As Soul ventures into the Great Void, Light moves into the great no thing. As it does so, parameters for new universes are laid out as magnetic filaments of Divine Light and Sound. Here, then, Soul has "begun" another eternity. Yet there is no beginning, for within that which has no thing, beginnings and endings are not possible. As Soul's greatest journey is undertaken—for it is Soul's task to co-create with Creator—That Which Is Omnipresent instantaneously becomes in that which knew not omnipresence. And so Divine Substance gives birth to nonsubstance.

When its ultimate task is complete, Soul returns Home.

Final Lesson, Earth Graduates

Dedicated in Love, our essence vibration instills peace and joy in the hearts of humans who are beginning to realize that the wheels of universal fortune have turned and that Earth's moment has come. She is about to be robed in prodigious stellar energies. Soon she will radiate naught but Love and Light. Representatives of many advanced civilizations have come from the stars to witness Earth's birth as a cosmic Christ planet. They have come to celebrate the triumphant reappearance of Prime Celestial Star-Master, the magnificent Dove, and to prepare their starseeded children for the journey Home.

Though you have presumed yourselves alone, Divine Christed Essence has always watched over you. You would surely be amazed to know the many times He has taken human form to remind you to Love all things unconditionally. It is so easy for humans to forget this simple but most profound of all spiritual teachings. It is so difficult for you to love that which appears unlovable.

Gentle Souls, you must remember that you are dearly beloved family to beings who hail from a multitude of stars. They come as a body of One to cherish and comfort you during turbulent times of massive change. They are pledged to assist your ascension efforts, to clothe you in Light—the standard attire in stellar regions where Love is the dominant feature.

Be aware, for as He has taken form during the most difficult times in human history, so will He still! Call His name as you will, be it Jesus, Buddha, Krishna, Mohammed, or Rama. It matters not. His I Am Presence lives as a tangible being within your hearts. He is contained in your awakening cellular Memories. His splendid Essence, long buried in your slumbering genetic filaments, is uncoiling like a ladder reaching for the stars. Do you hear Him calling you? Do you understand you are One of His own?

Within your lionlike exteriors lies a lamb. You are essentially gentle and loving beings. Even the most evil-appearing, in the depths of their core selves, are soft, sweet, and tenderly vulnerable. The absurd crusty shells you have deliberately put upon yourselves as you sought a false sense of security have kept you from experiencing yourselves as Soul. You are essential elements of a conscious, loving universe. Lifetime after lifetime spent in delusions of separation from God have erected formidable barricades of pain and suffering around you. Caught up in the depths of spiritual misery, you inadvertently thought you courted God's disfavor. Yet there is only one thing you have ever really longed for, and that is to reconnect with what you perceive as being severed.

Manifestation of I Am One lies within you as awakening aspects of Divine Soul. Your foundational makeup, as it is for all beings and all things, is Light. If you would view yourselves from a higher perspective you would see that the differences among humans are no differences. You would understand that every individual is a vital, unique facet in the tinkling crystalline entity that is uni-versal harmony. Your awakening auras sparkle like motes of stardust. Gender, race, or social status has no bearing on vibrational at-One-ment with Light. That which is believed to be dominant-superior or submissive-inferior is nothing more than an illusionary contrivance of third-dimensional beings.

We often refer to ourselves as primary aspects of Divine Everpresence—as is that which is you! In truth, there is no thing but Divine Presence eternally creating Itself. Thus, Oneness is capable of diversity of expression. Prolonged supposition otherwise has convinced humans that they must arrange themselves into a multitude of races and nations and take a determined stand against all others. This fearful attitude has made it possible for false leaders and false prophets to take dominance over you.

That which is Prime Leader is One. You, as are all beings, are One. You are the leader and you are the led. You are magnificent elohim reborn. That you Be! Aye!

Fortune blesses those who embrace peace and love as foundational heart-mind substance. Those of you who have the courage to explore the depths of your innermost being and nurture your tender feelings resemble splendid flowers coming to full bloom. You are ripe with abundance, and the seeds of light you sow will someday transform Earth's fertile lands into a reborn Eden.

Ecstasy of light becoming Light is beyond human comprehension. Your hearts are scorched with age-old sorrow from being caught in webs of karma indicative of Earth being the site of an ancient galactic battlefield. It is very difficult for you to grasp the brilliant vision of the future. When you close your eyes, accomplish this simple feat, and realize it is not so difficult after all, you will have taken a giant step in becoming One with beings who have mastered the stars. It is at this moment that you will consciously traverse the spatial-time bridge which joins us. This rainbow-hued bridgelike connector is actually an "elevator" among dimensions. It is that which draws you up and away from all categories of separatism. It is a vibrational transporter that moves from note to note, like scales played on a piano. You know that when you relax and reflect on beauty your thoughts move gently upward. Psychologically, this is well understood. What humans do not appreciate is that when they deliberately harmonize their emotions, they instigate an energy exchange that actually transports them from one vibration to another. This is a critical step in creating your own reality. What is not generally appreciated is that vibrational levels are actual space-time places.

Awakening starseeds are quite entranced with the notion of the higher Self. What they have not quite grasped, although they are coming closer, is that they are already higher Self. Higher Self is one and the same as that which is you. You are not dangled like a puppet by some outside-inside force. That which is you as a vital, living entity is already that which is you as Soul-Self.

Oneness is your final lesson, Earth graduates. One is what the great spiritual masters of yesteryear taught you. So said Christ. So said Buddha. So said Krishna. As was yesterday, so is today and so is tomorrow. Oneness is Omnipresent, Omniscient, Omnipotent Cosmic Intelligence as vibration. First there was the Word—Aum. As Divine Vibration interfaces with unconditional Love, cosmic Light takes form. As Divine Vibration and Light merge, universes and all they contain come into being.

Earth is a cosmic schoolhouse. You are the students. You are the teachers. Our role is to assist and advise you. As instructors our task is to present our students with an achievable lesson plan. If you are intent on acquiring a doctorate degree in science and wish to acquire higher knowledge of the nature of the universe, you must understand the nature of One. To be truly competent in any of the arts, you must comprehend the nature of One. This is equally important in any field of study, be it social sciences, history, philosophy, or the profoundly important area of mother/fatherhood. To master ascension, your final Soul-level proficiency test will be based on your knowledge of and ability to function as One.

Death is an illusion. You must let your fear of death go. As you let go of the illusion, you achieve a state of conscious grace. A state of grace—to be held in the wings of angels—is essential for light-body ascension. Acquisition of internal grace manifests in an unyielding desire to be done with separation, to be One in experience and expression. What humans have simply lost touch with is that they have always been One. Because this is a free-will universe, you have been allowed the illusion of separation for as long as you wished to keep it. Many worlds embraced within the structures of universal harmony nurture beings who wish to maintain a state of other-than. Historically, Earth has been one of those planets. Now, Earth is preparing for vibrational liftoff, and many humans have opted to join her. Magnificent beings of light stand ready to guide all who choose evolution's path.

Fashions allow much diversity of dress. Men, women, and children all delight in changes of clothing styles. If the material from which the cloth is made is of the best quality—laces, gabardines, silks, satins, or softly woven cottons—the sensation on the body is oh so fine. Encased in glamorous attire, proudly you strut down your streets with your heads held high. Symbolically, the donning of fine garments may be viewed as another aspect of the commonality that bonds you in Oneness. Your manner of dress is a reflection of yourself that you have chosen to demonstrate to others. It is one of the ways you say, "This is who I am." Your fascination with fine attire is a conscious symbolic illustration of your recollection of the garments of exquisite light that cover your Soul bodies—suits of liquid gold that sparkle like the brightest sun. It is that which you hold in common with all beings. It is your Spirit. It is your natural state of being.

Time broken into compartments called past, present, and future is also One. Higher beings experience no sense of separation from one experience to the next. As you move upward in resonation, you, too, will become immersed in the blissful state of now that is common to planetary beings who are swathed in Light. Besieged with Memories of what you thought had once been and hopeful or frightening visions of a future, it is not surprising that you are bogged down in spiritual confusion. In actuality, you have never been apart from the Moment of One. This is difficult for your curious, logic-loving intellects to comprehend. Eternal One presence is located in your hearts—where it has always been. For further edification, we suggest you study this universal concept from the vantage point of your awakening intuitive-mind.

This document on One represents a lesson plan that must be completed before you are granted graduation status and Gatekeepers allow you entry into higher realms.

Go in Peace, brothers and sisters of Earth.
Adonai.

Glossary

ADONAI: Hebrew for "Lord." Divine energy associated with the word's vibration assures Patricia of her light-level telepathic connection.

AKASHIC RECORDS: Cosmic journals that contain the records of the Soul's journey. These records are attended to by the angelic realms. With permission, Earth's spiritual masters and empowered enlightened beings are able to peruse the records.

ASCENDED MASTERS: Earth-incarnated souls who have overcome death, have assumed their bodies of light, and have attained God-realized Christ Consciousness.

ASTRAL PLANES: Fourth dimension, planes of instant manifestation. The astral is where reincarnating souls attached to Earth dwell between lives. The astral planes are vast and multilayered. The lower astral is where negative beings and negative thought forms reside. The region referred to in religious texts as heaven is the upper astral. See *Octave*.

AUM, OHM: Prime tone, a basic ingredient of universal energy. See *Hum.*

BLUE CRYSTAL PLANET: Closest English translation for the principal light-body planet of the Arcturian star system. Primary planning and gathering planet for the multiuniversal, multidimensional star councils.

BRAIN-MIND: The logical mind where the computerlike calculations of the physical brain are stored. See *Conscious mind, Heart-mind, Subconscious mind,* and *Superconscious mind.*

BROTHERHOODS OF LIGHT: *Brotherhood* means "in Oneness." There are many orders of brotherhoods of light. Earth's embodied and ascended masters are members of these brotherhoods. The Intergalactic Brotherhood of Light, which is made up of spiritually advanced extraterrestrials, is one order. Another important brotherhood associated with Earth's ascension is the Order of Melchizedek. The Office of the Christ heads the brotherhoods of light. *The Book of Knowledge: The Keys of Enoch* by J. J. Hurtak is an excellent resource for information on the brotherhoods. See "Suggested Books and Movies." See *Intergalactic Brotherhood of Light.*

CELESTIAL HOME: Also referred to as Central Sun. The Soul yearns to return to Celestial Home. It is the Soul's journey's end.

CELLULAR MATRICES: The molecular makeup of all third-dimensional physical matter. Used to describe the human body as well as Earth's body.

CHAKRA: Wheels of energy that make up the body's inner anatomy. Often described as lotus blossoms in Eastern tradition. Chakras are widely covered in Hindu, Buddhist, and Yoga texts. Shirley MacLaine's "Inner Workout" video is an excellent study and meditation source for the Western mind.

CHEUEL: An Arcturian sister planet to Earth destroyed by her citizens because of improper nuclear-energy use about five million years ago (in Earth historical time). Abundantly forested and populated with many animals and plants, it was recently restored by the Intergalactic Brotherhood to pristine form. It awaits Arcturian starseeds as a rest and recuperation planet after their missons on Earth are completed.

CONSCIOUS MIND: Third-dimensional brain functions. Mental layers of linear-logical thought. Same as brain-mind. See *Brain-mind, Heart-mind, Subconscious mind,* and *Superconscious mind.*

DARK LORDS: Evil, manipulative, controlling beings who throughout history have attempted to hold humanity in their clutches. They are referred to as satanic beings, Lucifer, and the dark angels. See *Grays.*

DENSITY OCTAVE, DIMENSIONAL OCTAVE: See *Octave.*

EAGLES OF THE NEW DAWN: Awakened humans (and animals) who interact with the star councils to serve Earth's evolution. Also called sky warriors.

ELOHIM: Ancient name for gods, also known in mythology as Zeus, Hercules, and so on. Those who manifest Light in accordance with Divine Plan. Now serve as galactic emissaries to Earth. See *Hybrid elohim.*

ENERGY FIELDS: Energy fields range from subtle to force-field magnitude. The energy field that surrounds the human body is the aura. See *Grids* and *Vortex*.

ETHERIC GRID STRANDS: See *Grids*.

FIFTH DIMENSION: Dimension of refined light. Arcturians are fifth- and sixth-dimensional beings. Negative beings are unable to penetrate into the realms of light substance.

FOURTH DIMENSION: See *Astral planes*.

GOD-REALIZED: An enlightened, evolved individual who has attained Christ Consciousness. Spiritual masters are God-realized.

GRACE: To receive grace is to be held in the fluttering arms of angels. Grace is an energy field that descends upon humans from Divine sources for physical, emotional, and mental healing, and to assist in fulfilling the Soul's life purpose.

GLOSSARY

GRAYS: Manipulative extraterrestrials who are in alliance with the Dark Lords. The Grays are responsible for human abductions and cattle mutilations. See *Dark Lords*. See *Hidden Mysteries* by Joshua D. Stone and *UFOs and the Nature of Reality* from Ramtha for detailed information.

GRIDS, GRIDLINES, SPACE GRIDS, STARGRIDS, STRANDS: Crisscrossing webs of light, sound, color, and scent that make up Earth's spiritual body. Starships also travel upon grids of light that weave through space. Grids of light connect galaxy to galaxy, star to star, planet to planet, moon to moon, and so forth. See Volume I, Part III for more detailed information.

HARMONIC COORDINATES OF THE GALACTIC HUM: The sound coordinates of the grids. See *Hum*.

HEART-MIND: The intuitive, spiritual mind. The heart chakra is the location of the heart-mind. It is where we connect with our higher Selves and our Soul Memories. See *Brain-mind, Conscious mind, Subconscious mind,* and *Superconscious mind*.

HUM: Cosmic creative vibration, the prime or God energy expressed as Aum (Amen) or Ohm. Matter's tonal qualities. Pythagoras described hum as the music of the spheres.

HYBRID ELOHIM: Human starseeds that maintain God-Soul Oneness at superconscious level; specific elements within a segment of starseeds who came to Earth from highly evolved planetary systems. See *Elohim*.

I AM: The Self's identity as Soul.

INTERGALACTIC BROTHERHOOD OF LIGHT: Spiritually evolved, light-formed, fifth- and sixth-dimensional extraterrestrials from many star systems and many universes. See *Spiritual Hierarchy* and *Star councils*.

LANGUAGE OF THE SUN: Common mode of telepathic communication natural to all beings. Also called solar tongue or solar language.

LIGHT-LOVE OR LOVE-LIGHT: Light is the first manifestation of God in form; Love is God's Essence or energy. Light-Love incorporates Creation's energy as a physical manifestation. Sound (vibration) is integral to Light-Love energy. Throughout the ages Earth's

God-realized spiritual masters have referred to Light-Love as unconditional Love.

LIGHT STRANDS: See *Grids*.

LORDS OF DARKNESS: See *Dark Lords*.

LOVE AND LOVE: Unconditional Love is integrated Cosmic Intelligence. However, love is the emotion humans feel for others, their pets, and Earth.

MANITU: Meaning "spirit keeper," it is the title the Intergalactic Brotherhood bestows on people whose life's purpose is planetary healing.

MARIGOLD–CITY OF LIGHTS: Intergalactic Brotherhood Earth-based mother ship. Authors may refer to it by other names, perhaps *City of Lights*, *Jeweled City*, *Crystal City*, *New Jerusalem*. The term *Marigold* was given to this writer as a symbolic clue that cosmic light incorporates vibration, color, and scent. The mother ship in the movie *Close Encounters of the Third Kind* portrays her essence, though *Marigold–City of Lights* is much larger.

MEMORIES: An aspect of Self-knowledge that slumbering humanity has forgotten; Soul Memories. As we awaken to our spiritual nature, the Memories are reactivated.

MERKABA (MER-KA-BA): A Star of David–shaped force-field of unconditional Love-Light that surrounds a starseed's awakening etheric body and, indeed, is the makeup of the higher-dimensional light-body. *The Book of Knowledge: The Keys of Enoch* by J. J. Hurtak describes the merkaba as a divine-light vehicle. Higher-dimensional starships are merkaba-light manifestations.

OCTAVE: A dimensional or density span. The vibrational layers within a dimension are not unlike the notes of the musical scale, ever softening in an upward or refining manner.

ORDER OF MELCHIZEDEK: Also known as Brotherhood of Melchizedek. One of the primary orders of the Brotherhoods of Light. In cooperation with the Office of the Christ, dispensers of Divine Light throughout this galactic sector. Sanat Kumara is the head Kumara of the Order. See Part II. See *The Book of Knowledge: The Keys of Enoch* by J. J. Hurtak for in-depth information.

PSYCHIC PROTECTION: The purposeful use of Light-Love when meditating or channeling. The following steps are recommended: cleansing one's physical environment with incense or sacred herbs (smudging); requesting Christ Consciousness energy (see *Spiritual Hierarchy*) to assist; visualizing light running through the chakras and surrounding the body in an energy bubble; using a repetitive, vibrational tone or mantra (for example, Aum or Ohm); and routinely forcefully challenging any entity with the statement "Are you of the Light (spiritual beings)?" (See Volume I, Part II, "Patricia Meets Palpae.") Beings of light expect to be challenged. Negative beings cannot penetrate the layers of a Light-Love established force-field.

RESONANT VIBRATIONAL HUM: Degree of refined light vibrating within a dimensional octave.

SIXTH DIMENSION: See *Fifth dimension.*

SKY WARRIORS: See *Eagles of the new dawn.*

SPACE GRIDS: See *Grids.*

SPACE POPPIES: Refers to scented, flowered coordinates that are the by-products or the bouquets of the harmonics. The Arcturians find the symbolic use of flowers helpful in explaining the fragrant qualities of the light grids.

SPACE-TIME: In this text, refers to the linear structures of third-dimensional spatial and time realities as well as to the ebb and flow of fourth-dimensional space and time.

SPIRITUAL HIERARCHY: Body of One, supreme spiritual council in service to Earth's ascension. The central figure is the Christ Essence. The council includes archangels and angelic realms, ascended masters, the brotherhoods of light (including the Intergalactic Brotherhood of Light), and God-realized humans.

STAR COUNCILS: Coordinators and directors of multi-spatial, intergalactic business affairs. All aware galactic citizens have input upon the star councils. Arcturians in service to Earth sit upon the star councils as a subdivision of the Supreme Hierarchical Council for Planetary Ascension, System Sol, Intergalactic Brotherhood of Light.

STARGRIDS: See *Grids*.

STARGATE: A multidimensional access window. See the movies *2001: A Space Odyssey* and *Stargate* for a graphic portrayal of stargate dynamics.

STARSEEDS: Galactic beings on Earth as humans, animals, and plants. Many universes and star systems are represented, among them Arcturus, Pleiades, Sirius, and Orion.

STRANDS: See *Grids*.

SUBCONSCIOUS MIND: The spirit-mind, or intuitive heart-mind, that receives information from the super-conscious, transforms it into symbols the brain-mind can understand, and then sends it to the brain-mind, or conscious mind. See *Brain-mind, Conscious mind, Heart-mind,* and *Superconscious mind.*

SUPERCONSCIOUS MIND: Soul mind, or higher-Self mind. It uses the subconscious mind as a conduit to send information to the brain-mind, or conscious mind. See *Brain-mind, Conscious mind, Heart-mind,* and *Superconscious mind.*

TITANIC BEINGS: Extremely advanced energies, propellants that dominate the creative urges of the universe and bring into form galaxies, stars, and planets. Currently engaged in driving evolved universal contents into the arms of Omnipresent, Omniscient Source.

UNI-VERSE/UNI-VERSAL: One Song. One Verse. One.

UNIVERSAL ONENESS COUNCIL OF TWENTY-FOUR ELDERS: Responsible for dispensing Primary Light throughout this galaxy. Composed of beings who blend and bond light and sound into birthing and advancing star systems.

UNIVERSAL LAW: The Law of One. One song (for) all.

VORTEX: Varying degrees of heightened energies that arise from a point along a grid where light strands crisscross. Vibration and light energy arising from a vortex varies greatly and may range from a few inches to miles. Intuitively, humans have always recognized vortices as power spots or sacred sites. See Volume I, Part III for more detailed information.

YIN-YANG: Ancient Chinese symbol used in the teachings of the Tao and the I Ching. Yin is feminine; yang is masculine. Yin-yang demonstrates all polarities and diversities (the ten thousand things) that exist within the universal whole.

Suggested Books and Movies

BOOKS

Agartha: A Journey to the Stars. Meredith Lady Young Sowers. Walpole, N.H.: Stillpoint, 1984.

Aliens Among Us. Ruth Shick Montgomery. New York: Fawcett, Crest, 1985.

An Act of Faith: Transmissions from the Pleiades. P'taah (Spirit), [channelled by] Jani King. The P'taah Tapes series. Cairns, Queensland, Australia: Triad, 1991; York Beach, Maine: Samuel Weiser, 1996.

Ancient America. Jonathan Norton Leonard. Great Ages of Man; A History of the World's Cultures series. (Nazca Lines.) New York: Time-Life Books, 1967.

Arcturus Probe. José Argüelles. Sedona, Ariz.: Light Technology Communication Services, 1996.

Autobiography of a Yogi. Paramahansa Yogananda. Los Angeles: Self-Realization Fellowship, 1946.

Bashar: Blueprint for Change: A Message from Our Future. Bashar (Spirit), [channelled by] Darryl Anka. Edited by Luana Ewing. Seattle: New Solutions, 1990.

Beyond Ascension. Joshua D. Stone. Sedona, Ariz.: Light Technology Communication Services, 1995.

Beyond Stonehenge. Gerald S. Hawkins. New York: Harper & Row, 1973; New York: Marboro Books, Dorset, 1989.

The Book of Knowledge: The Keys of Enoch. J. J. Hurtak. Los Gatos, Calif.: Academy for Future Science, 1977.

Bringers of the Dawn: Teachings from the Pleiadians. Barbara Marciniak. Santa Fe, N.M.: Bear, 1992.

Celestial Raise: 'Tiers of Light' Pouring Fourth from the Son. Edited by Marcus. Mt. Shasta, Calif.: ASSK (Association of Sananda and Sanat Kumara), 1986.

The Chinese Roswell: UFO Encounters in the Far East from Ancient Times to the Present. Hartwig Hausdorf. Translated from the German by Evamarie Mathaey and Waltraut Smith. Boca Raton, Fla.: New Paradigm Books, 1998.

Circular Evidence: A Detailed Investigation of the Flattened Swirled Crops Phenomenon. Pat Delgado and Colin Andrews. Grand Rapids, Mich.: Phanes Press, 1989.

The Complete Ascension Manual for the Aquarian Age. Joshua D. Stone. Sedona, Ariz.: Light Technology Communication Services, 1994.

Connecting with the Arcturians. [Channelled by] David K. Miller. Pine, Ariz.: Planetary Heart Publications, 1998.

Conversations with Eternity: The Forgotten Masterpiece of Victor Hugo. Translated with a commentary by John Chambers. Boca Raton, Fla.: New Paradigm Books, 1998.

The Cosmic Connection: Worldwide Crop Formations and ET Contacts. Michael Hesemann. Bath, England: Gateway Books, 1996.

The Crystal Stair: A Guide to the Ascension: Channeled Messages from Sananda (Jesus), Ashtar, Archangel Michael, and St. Germain. Eric Klein. Edited by Sara Benjamin-Rhodes. Livermore, Calif.: Oughten House, 1990; third edition, 1994.

The Divine Romance. Paramahansa Yogananda. Los Angeles: Self-Realization Fellowship, 1986.

The Earth Chronicles series. Vols. I–V. Zecharia Sitchin. New York: Avon Books; Santa Fe, N.M.: Bear, 1980–1993.

Earth's Birth Changes. St. Germain (Spirit), [channelled by] Azena Ramada. St. Germain Series. Cairns, Queensland, Australia: Triad; York Beach, Maine: Samuel Weiser, 1996.

E.T. 101: The Cosmic Instruction Manual for Planetary Evolution. Mission Control (Spirit), [channelled by] Zoev Jho. San Francisco: HarperSanFrancisco, 1994. Originally published as channelled by Diana Luppi (1990).

Explorer Race Series. Vols. I–VI. Zoosh (Spirit), [channelled by] Robert Shapiro. Sedona, Ariz.: Light Technology Publishing, 1996–1998.

Family of Light. Barbara Marciniak. Santa Fe, N.M.: Bear, 1999.

The Findhorn Garden. Findhorn Foundation. New York: HarperCollins, 1975.

God I Am: Inspired by the Triad of Isis, Immanuel and St. Germain. Peter O. Erbe. Cairns, Queensland, Australia: Triad; York Beach, Maine: Samuel Weiser, 1996.

The Gods of Eden: A New Look at Human History. William Bramley. New York: Avon Books, 1989.

Hathor Material. Tom Kenyon and Virginia Essene. Santa Clara, Calif.: S.E.E. (Spiritual Education Endeavors) Publishing Co., 1996.

Hidden Mysteries. Joshua D. Stone. Sedona, Ariz.: Light Technology Communication Services, 1995.

Kryon Series. Vols. I–VI. Kryon (Spirit), [channelled by] Lee Carroll. Del Mar, Calif.: Kryon Writings, 1993–1997.

Lazaris (Spirit) series books, videos, and cassettes. Palm Beach, Fla.: Visionary Publishing.

Life and Teachings of the Masters of the Far East. Vols. I–VI. Baird T. Spalding. Marina del Rey, Calif.: DeVorss, 1924.

Mary's Message to the World: As Sent by Mary, the Mother of Jesus, to Her Messenger Annie Kirkwood. Mary, Blessed Virgin, Saint (Spirit), [channelled by] Annie Kirkwood. Edited by Brian Kirkwood. New York: Putnam, 1991; New York: Berkley, Perigee, 1996.

The Mayan Factor: Path Beyond Technology. José Argüelles. Santa Fe, N.M.: Bear, 1987.

The Monuments of Mars: A City on the Edge of Forever (book and video). Richard C. Hoagland. Berkeley, Calif.: North Atlantic Books, Frog, Ltd., 1987.

The Nature of Personal Reality: Specific, Practical Techniques for Solving Everyday Problems and Enriching the Life You Know (and other Seth series books). Seth (Spirit), [channelled by] Jane Roberts. Englewood Cliffs, N.J.: Prentice-Hall, 1974; San Rafael, Calif.: Amber-Allen, 1994.

Nothing in This Book Is True, But It's Exactly How Things Are: The Esoteric Meaning of the Monuments on Mars. Bob Frissell. Berkeley, Calif.: North Atlantic Books, Frog, Ltd., 1994.

The Only Planet of Choice: Essential Briefings from Deep Space. Phyllis V. Schlemmer. Edited by Mary Bennett. Bath, England: Gateway Books, 1993; revised edition, 1996.

A Path with Heart: A Guide through the Perils and Promises of Spiritual Life. Jack Kornfield. New York: Bantam Books, 1993.

The Pleiadian Agenda: A New Cosmology for the Age of Light. Barbara Hand Clow. Santa Fe, N.M.: Bear, 1995.

Project World Evacuation: UFOs to Assist in the "Great Exodus" of Human Souls off this Planet. Compiled through Tuella by the Ashtar Command. Edited by Timothy Green Beckley. Petaluma, Calif.: Inner Light Publications, 1993.

Ramtha. Ramtha (Spirit), [channelled by] J. Z. Knight. Edited by Steven L. Weinberg. Eastsound, Wash.: Sovereignty, 1986.

The Sirius Connection: Unlocking the Secrets of Ancient Egypt. Murry Hope. Rockpart, Mass.: Element Books, 1996.

Something in This Book Is True–. Bob Frissell. Berkeley, Calif.: North Atlantic Books, Frog, Ltd., 1997.

The Star-Borne: A Remembrance for the Awakened Ones. Solara. Charlottesville, Va.: Star-Borne, 1989.

The Starseed Transmissions. Ken Carey. San Francisco: HarperSanFrancisco, 1991.

Surfers of the Zuvuya: Tales of Interdimensional Travel. José Argüelles. Santa Fe, N.M.: Bear, 1988.

The Third Millennium: Living in the Posthistoric World. Ken Carey. San Francisco: HarperSanFrancisco, 1995. Originally published as *Starseed, the Third Millennium* (1991).

The Tibetan Book of Living and Dying: A New Spiritual Classic from One of the Foremost Interpreters of Tibetan Buddhism to the West. Sogyal Rinpoche. San Francisco: HarperSanFrancisco, 1993.

Transformation of the Species: Transmissions from the Pleiades. Jani King. The P'taah Tapes series. Cairns, Queensland, Australia: Triad.

The Treasure of El Dorado: Featuring "the Dawn Breakers." Joseph Whitfield. Roanoke, Va.: Treasure, 1977; reprint, 1989.

UFOs and the Nature of Reality: Understanding Alien Consciousness and Interdimensional Mind. Ramtha (Spirit), [channelled by] J. Z. Knight. Edited by Judi Pope Koteen. Eastsound, Wash.: Indelible Ink, 1990.

We, the Arcturians. Norma J. Milanovich. Albuquerque, N.M.: Athena, 1990.

With Wings As Eagles: Discovering the Master Teacher in the Secret School Within. John R. Price. Boerne, Texas: Quartus Books, 1987; Carson, Calif.: Hay House, 1997.

You Are Becoming a Galactic Human. Washta (Spirit), [channelled by] Virginia Essene and Sheldon Nidle. Santa Clara, Calif.: S.E.E. (Spiritual Education Endeavors) Publishing Co., 1994.

MOVIES ("MANAGEMENT TRAINING FILMS")

2001: A Space Odyssey. Stanley Kubrick film.

2010: The Year We Make Contact. Peter Hyams film.

Abyss. James Cameron film.

Always. Steven Spielberg film.

Batteries Not Included. Steven Spielberg film.

City of Angels. Brad Silberling film.

Close Encounters of the Third Kind (extended version). Steven Spielberg film.

Cocoon. Ron Howard film.

Contact. Robert Zemeckis film.

Defending Your Life. Geffen Pictures.

Field of Dreams. P. A. Robinson film.

Ghost. Jerry Zucker film.

Heart and Souls. Ron Underwood film.

Heaven Can Wait. Paramount.

Hoagland's Mars. Richard C. Hoagland.

Made in Heaven. Lorimar Motion Pictures.

Phenomenon. Touchstone Pictures.

Planetary Traveler. Third Planet Entertainment.

Powder. Hollywood Pictures.

Shirley MacLaine's Inner Workout. High Ridge Productions.

Star Trek (entire series, both motion pictures and television, especially *Star Trek IV: The Voyage Home*).

Star Wars trilogy: *Star Wars, The Empire Strikes Back, Return of the Jedi, The Phantom Menace.* George Lucus films.

Stargate. Mario Kassar film.

Starman. John Carpenter film.

What Dreams May Come. Polygram Films.

Willow. Ron Howard film.

THE ARCTURIAN STAR CHRONICLES SERIES
Patricia L. Pereira started receiving telepathic communica-
tions from the star Arcturus in 1987 and transcribed a message of hope and encouragement about the changes we will
experience in the years to come. These galactically inspired
pages have become the Arcturian Star Chronicles.

Volume One
Songs of the Arcturians
$12.95 softcover
Practical and uplifting cosmically inspired manual
designed to assist the individual in preparing for galactic
citizenship and matters pertaining to personal evolution.

Volume Two
Eagles of the New Dawn
$12.95 softcover
Galactic essays and exercises to assist awakening humans
(eagles of the new dawn) in unlocking their soul memories
and purposefully connecting with higher-dimensional
spirit energies.

Volume Three
Songs of Malantor
$13.95 softcover
Cosmic information of expanded complexity to assist
humans in times of change and to prepare them for
citizenship in the greater galactic community.

Volume Four
Arcturian Songs of the Masters of Light
$13.95 softcover
Moves beyond solely Arcturian energies and introduces
the Titanic Beings of the Galactic Core.

These titles are available through your local bookstore
or from Beyond Words Publishing at 1-800-284-9673.

Beyond Words Publishing, Inc.

OUR CORPORATE MISSION:
Inspire to Integrity

OUR DECLARED VALUES:
We give to all of life as life has given us.
We honor all relationships.
Trust and stewardship are integral to fulfilling dreams.
Collaboration is essential to create miracles.
Creativity and aesthetics nourish the soul.
Unlimited thinking is fundamental.
Living your passion is vital.
Joy and humor open our hearts to growth.
It is important to remind ourselves of love.